# When Topology Meets Chemistry

The applications of topological techniques for understanding molecular structures have become increasingly important over the past thirty years. In this topology text, the reader will learn about knot theory, three-dimensional manifolds, and the topology of embedded graphs, while learning the role these play in understanding molecular structures. Most of the results that are described in the text are motivated by questions asked by chemists or molecular biologists, though the results themselves often go beyond answering the original question asked. There is no specific mathematical or chemical prerequisite; all the relevant background is provided. The text is enhanced by nearly 200 illustrations and more than 100 exercises.

Reading this fascinating book, undergraduate mathematics students can escape the world of pure abstract theory and enter that of real molecules, whereas chemists and biologists will find simple and clear but rigorous definitions of mathematical concepts they handle intuitively in their work.

Erica Flapan received her PhD from University of Wisconsin in 1983 in the field of knot theory. She held positions at Rice University and the University of California at Santa Barbara before joining the faculty at Pomona College, where she is currently a professor. She became interested in the applications of topology to chemistry in 1984, after learning about Jon Simon's proof that molecular Möbius ladders were topologically chiral. In addition to her many publications on three-dimensional manifolds and knot theory, she has published numerous papers in mathematics and chemistry journals about graphs embedded in space and their applications to chemistry.

# Outlooks

Published by Cambridge University Press and
the Mathematical Association of America

Mathematical content is not confined to mathematics. Eugene Wigner noted the unreasonable effectiveness of mathematics in the physical sciences. Deep mathematical structures also exist in areas as diverse as genetics and art, finance, and music. The discovery of these mathematical structures has in turn inspired new questions within pure mathematics.

In the Outlooks series, the interplay between mathematics and other disciplines is explored. Authors reveal mathematical content, limitations, and new questions arising from this interplay, providing a provocative and novel view for mathematicians, and for others an advertisement for the mathematical outlook.

# When Topology Meets Chemistry

## A Topological Look
## at Molecular Chirality

Erica Flapan

*Pomona College*

MATHEMATICAL ASSOCIATION OF AMERICA

CAMBRIDGE
UNIVERSITY PRESS

CAMBRIDGE UNIVERSITY PRESS
Cambridge, New York, Melbourne, Madrid, Cape Town, Singapore,
São Paulo, Delhi, Dubai, Tokyo

Cambridge University Press
32 Avenue of the Americas, New York, NY 10013-2473, USA

www.cambridge.org
Information on this title: www.cambridge.org/9780521664820

MATHEMATICAL ASSOCIATION OF AMERICA
1529 Eighteenth St. NW, Washington DC 20036    www.maa.org

First published 2000
Reprinted 2007

*A catalog record for this publication is available from the British Library*

*Library of Congress Cataloging in Publication data*
Flapan, Erica, 1956–
When topology meets chemistry / Erica Flapan.
p.   cm. – (Outlooks)
Includes bibliographical references and index.
ISBN 0-521-66254-0 (hb) – ISBN 0-521-66482-9 (pb)
1. Chemistry, Physical and theoretical – Mathematics.   2. Chirality.   3. Topology.   4. Knot
theory.   I. Title.   II. Series.
QD455.3.T65 F53   2000
541'.01'514 – dc21        00-027517

ISBN 978-0-521-66254-3 Hardback
ISBN 978-0-521-66482-0 Paperback

Transferred to digital printing 2009

To Francis and Laure

# Contents

# Preface

The fields of topology, chemistry, and molecular biology have each made tremendous advances in the past thirty years, and as a result, there are more and more potential ways that topological techniques can be used to understand molecular structures. In particular, there have recently been a number of conferences and special issues of journals devoted to research that is at the intersection of topology, chemistry, and molecular biology. Although this work is all quite exciting, it is often technically difficult to wade through for those outside of these specific interdisciplinary areas. With some effort, mathematicians are able to bring together various sources and make sense of these results. In my experience teaching undergraduates, I have found that topology students are frequently eager to learn about results in topology that can be applied to science; however, they have quite a bit of difficulty understanding the existing literature. Many chemists and molecular biologists would also like to learn about topological techniques that they could use. However, scientists generally do not have the topological background necessary to read these results carefully. In addition to research articles, expository versions of some of these topics have been published with the details omitted. Although this has enabled many people to get a sense of the types of results that have been obtained, there has been no topology text that is written at an appropriate level for mathematics students and scientists who want to learn the topology that is necessary to understand the details of the interdisciplinary work.

This book is an attempt to provide such a text. It is a topology book written in the context of how topology can be applied to chemistry and molecular biology. The reader will learn a good deal of low dimensional topology, knot theory, and the topology of embedded graphs, while learning how these concepts can play a role in understanding molecular structures. Many of the topological results that I will describe were originally motivated by questions asked by chemists or molecular biologists, though the topological results themselves often go beyond answering the specific question asked. In fact, in some cases, questions arising in the scientific community have led to topological results that are of interest in their own right. Although not all of this work will be tied directly to chemistry or biology, I believe that all of it has at least potential applications to these fields.

This book is an expanded version of a series of lectures that I gave at the Centre Emile Borel while participating in a special semester on low dimensional topology in the spring of 1996. The audience for those lectures was primarily mathematics graduate students. However, my lectures were presented at a level that was accessible to advanced undergraduates without much background in topology. In writing up the notes as a manuscript, I have tried to make much of the material accessible to chemists and molecular biologists as well as to students of mathematics. The text assumes that the reader has a basic understanding of continuous functions of Euclidean space, but it has no other specific prerequisites in mathematics or chemistry. I have provided definitions for virtually all topological terms and tried to keep the book as nontechnical as possible. Students with some background in geometric topology should have no trouble understanding the book in complete detail. Those without such a background might have some difficulty with the proofs of Theorems 6.1 and 7.1, which make use of some deep results in topology. These two proofs are sketched rather than presented in complete detail, and those who are uncomfortable with the material should feel free to skip or skim these proofs. I include these two proofs in order to expose the reader to the power of current topological machinery, with the hope that this will give the reader some motivation to continue studying topology. The proofs of all of the results other than Theorems 6.1 and 7.1 can be understood based on the material developed in the text. Those readers who wish to read the concepts and examples without delving into many of the proofs should note that the end of each proof is denoted by a box □, so they can easily skip over whichever proofs they choose.

There are exercises at the end of each chapter that are meant to reinforce the definitions and proofs provided in the text. Many of the exercises can be done either intuitively or more rigorously, depending on the mathematical background of the reader.

All of the topological spaces considered in this book are actually subsets of Euclidean $n$-dimensional space. So, in order to avoid unnecessary abstraction, I never actually use the terms *topological space* or even *metric space*. Rather, I work exclusively with the usual Euclidean metric. In addition, I have not specified when I am working in the piecewise linear category and when I am working in the differentiable category. This should not concern most readers. However, those readers concerned about such things should assume that any homeomorphism is either piecewise linear or is a diffeomorphism according to whichever one is more appropriate in the given context.

I thank the Centre Emile Borel for inviting me to participate in the special semester on low dimensional topology and giving me the opportunity to present these lectures. I thank my former students Brian Forcum, Helgi Bloom, and Elizabeth John for creating many of the drawings. Thanks also go to Corinne Cerf for reading an early version of the manuscript and making suggestions on how it could be improved. I also thank Asuman Aksoy for suggesting the title of the book. Finally, I thank Kurt Mislow, whose papers on Chemistry and Topology have stimulated my interest and many of my ideas.

# 1

# Stereochemical Topology

Stereochemistry is the study of the three-dimensional structure of molecules, and topology is the study of those properties of geometric objects that are invariant under continuous transformations. It is not obvious that these two fields have anything in common. In fact, not long ago there was little communication between researchers in these two areas. Prior to forty years ago, analyzing the topological properties of existing molecular structures was not very difficult, because as topological objects, the graphs of all the molecular structures known at the time could be deformed into a plane. Thus, understanding the stereochemistry of a molecule only required the evaluation of its geometry and not its topology. Recently, knots and links and other molecules have been synthesized whose structures and properties come from their topology as well as their geometry. The chemical motivation for the synthesis of such topologically interesting structures is the desire to synthesize new types of molecules that might have truly unusual properties, as well as the hope that in trying to create such unique molecules, new methods of synthesis will be developed along the way. These new types of molecules are often large enough that they no longer have the rigidity that is characteristic of small molecules, so understanding their deformations is an important part of understanding their structure. In addition to purely synthetic molecules, Liang and Mislow (1994b, 1995) have discovered that knots and links and other nonplanar graphs can occur naturally in proteins. Also, molecular biologists have found that knotted and linked DNA can exist in nature (Clayton & Vinograd, 1967; Hudson & Vinograd, 1967; Liu, Depew, & Wang, 1976) and can be manipulated in the lab (Seeman, 1999). Thus now both chemistry and molecular biology have something to gain by understanding the topology of graphs embedded in three-dimensional space.

Although many applications of topology to chemistry have recently gained attention, it is worth noting that chemistry also made an important contribution to topology over a century ago in the formation of the field of knot theory.

1

In the 1880s, scientists believed that all of space was filled with a substance called *ether*. William Thomson (a.k.a. Lord Kelvin) theorized that atoms were vortex rings in the ether; rings that were knotted and linked in different ways represented different elements (Kelvin, 1904). In order to try to create a periodic table of the elements, the physicist Peter Guthrie Tait created tables that divided knots into classes in such a way that a knot in one class could not be deformed into a knot in another class (Tait, 1898). By the turn of the century, it was clear that Kelvin's theory of atomic structure was wrong. Thus chemists were no longer interested in classifying knots. However, topologists had become interested in the study of knotted closed curves in space, and the field of knot theory has blossomed since then into an important subfield of three-dimensional topology that is closely tied to many other areas of topology. As we shall see, many of the ideas of knot theory are now the basis for some of the applications of topology to stereochemistry.

This chapter provides an overview of some of the topological theorems that have been proven, as well as some of the open problems that have arisen as a result of applying topological techniques to the study of complex molecular structures. Rigorous definitions or proofs are not given in this chapter; rather the focus is on providing a visual and conceptual understanding. The results that are mentioned here, as well as other results, will be proven in detail in Chapters 2 through 6. Chapter 7 is devoted entirely to the topology of DNA.

## Examples of Topologically Complex Molecules

In this section we will give a very brief overview of some of the topologically complex molecules that have been synthesized. By a *graph* we shall mean a finite collection of vertices together with disjoint edges connecting pairs of vertices, with the requirement that there is at most one edge between any pair of vertices and every edge has two distinct vertices. Figure 1.1 illustrates the types of edges that are forbidden in a graph. An *embedded graph* is a graph that exists in a specific position in three-dimensional space, whereas an *abstract graph* is a graph that is considered to be independent of any particular embedding in three-dimensional space. The *molecular bond graph* of a molecule is a model of the molecule, which is represented by a graph embedded in three-dimensional space where the vertices represent atoms or collections of atoms and the edges represent chemical bonds or chains of atoms. Such a molecular bond graph is often labeled to indicate what atom or atoms each vertex represents, and to indicate which edges represent double bonds and which represent single bonds.

A graph in three-dimensional space is considered to be topologically complex if it cannot be deformed into a plane, even under the assumption of complete

**Figure 1.1.** A graph is not permitted to have these types of edges

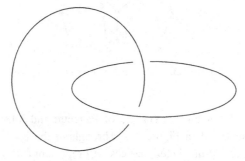

**Figure 1.2.** The first molecular link to be synthesized had the form of a Hopf link

flexibility. If we imagine that the graph of a typical molecule is made out of rubber, it can easily be deformed into a plane. Of course many molecules are actually too rigid to be deformed into a plane. Nonetheless, any structure that could theoretically be deformed into a plane will not be the object of our study.

The first topologically complex molecules to be synthesized were a pair of linked rings, known in chemistry literature as *catenanes*. Although the individual rings of a catenane are disjoint, they behave together as a single molecule. Starting around 1910, chemists began trying to synthesize a pair of linked rings. However, a molecular ring had to be synthesized that was big enough so that another molecule could pass through the hole in the center in order to create a link. Synthesizing a catenane, and proving that it was such, was finally achieved in 1961 by Frisch and Wasserman (Frisch & Wasserman, 1961). Their catenane was created out of a pair of linked hydrocarbon rings that each contained 34 atoms. A *hydrocarbon* is a compound that is made up entirely of hydrogen and carbon atoms. This molecular link has the structure of the *Hopf link*, which is illustrated in Figure 1.2. This is the simplest possible link, as it is the only link that can be drawn with only two crossings. Even if we imagine that the molecular graph of this link is completely flexible, it cannot be deformed into a plane because the rings are linked and cannot be separated unless one is cut. Although Frisch and Wasserman discussed the possible synthesis of knots, they did not succeed in synthesizing a knot.

Another topologically important molecule is the Simmons–Paquette molecule, which was independently synthesized in 1981 by the laboratories

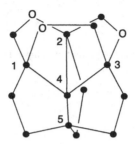

**Figure 1.3.** The Simmons-Paquette molecule

of Simmons and Maggio (1981) and of Paquette and Vazeux (1981). This molecule is illustrated in Figure 1.3. Throughout the book, when we draw molecular graphs, all unlabeled vertices will represent carbon atoms, and for the sake of simplicity, hydrogen atoms will generally be omitted. In Figure 1.3, the three Os represent oxygen atoms. The numbers on some of the vertices have no chemical significance and will be explained shortly. Like that of the linked rings, even if it were completely flexible, the Simmons–Paquette molecular graph could not be deformed into a plane. However, this graph has the even stronger property that no embedding of the graph can lie in a plane. Thus there is no way to take apart the graph of the Simmons–Paquette molecule into a collection of vertices and edges and reassemble the vertices and edges in a plane in such a way that edges connect the same pairs of vertices as were connected in the original molecular graph and no edges intersect except at their vertices.

A graph with the property that no embedding of it can lie in a plane is said to be *nonplanar*. Nonplanarity is a property of an abstract graph, rather than of the specific way that the graph is embedded in three-dimensional space. For example, the Hopf link, with any number of vertices added to each ring, is a specific embedding of a planar graph, although the particular embedding illustrated in Figure 1.2 cannot be deformed into a plane. We could take apart the vertices and edges and reassemble them as two disjoint circles in the plane. Kuratowski (1930) proved that a graph is nonplanar if and only if it contains either the complete graph on five vertices, $K_5$, or the bipartite graph on three vertices, $K_{3,3}$, each with the possible addition of some extra vertices. We illustrate the graphs $K_{3,3}$ and $K_5$ in Figure 1.4. We have drawn these graphs abstractly, without specifying how they are embedded in space. In particular, these pictures are not meant to indicate that the edges intersect. Rather, no matter how these (or any other) graphs are positioned in three-dimensional space, their edges are required to be disjoint except at the vertices.

A *complete graph* on $n$ vertices, $K_n$, consists of $n$ vertices together with edges connecting every pair of vertices. A *complete bipartite graph*, $K_{p,q}$, consists

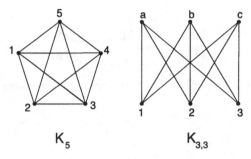

**Figure 1.4.** Any nonplanar graph contains either $K_{3,3}$ or $K_5$

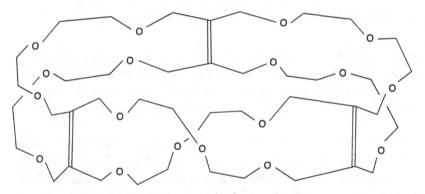

**Figure 1.5.** A molecular Möbius ladder

of a set of $p$ vertices and a set of $q$ vertices together with edges connecting every vertex in one set to every vertex in the other set. We have numbered five of the vertices in Figure 1.3 to indicate that the Simmons–Paquette graph actually contains a $K_5$. The reader should observe that in the Simmons–Paquette molecular graph, there are disjoint paths connecting any numbered vertex to any other numbered vertex. Thus if we were to omit those vertices of the Simmons–Paquette without numbers, we would obtain a complete graph on five vertices. Hence it follows from Kuratowski's theorem that the Simmons–Paquette graph cannot be embedded in a plane.

In 1982, Walba, Richards, and Haltiwanger (Walba et al., 1982) synthesized the first molecular Möbius ladder with three rungs (see Figure 1.5). This molecule resembles a Möbius strip in which the surface of the strip has been replaced by a ladder. The sides of the ladder represent a polyether chain of 60 atoms that are all carbons and oxygens, and the rungs of the ladder represent carbon–carbon double bonds. Thus the type of edges represented by the sides are chemically quite different from those represented by the rungs. We can see that

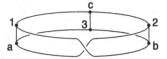

**Figure 1.6.** A Möbius ladder with three rungs is an embedding of $K_{3,3}$

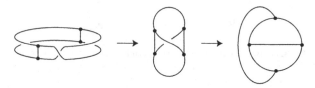

**Figure 1.7.** A two-rung Möbius ladder can be deformed into a plane

the three-rung Möbius ladder is a nonplanar graph because it contains a $K_{3,3}$ graph. Figure 1.6 represents the structure of a Möbius ladder with three rungs, where the vertices of valence 2 have been omitted and the remaining vertices have been labeled 1, 2, 3 and $a$, $b$, $c$ in order to help the reader see that this graph is a $K_{3,3}$.

In 1975, a two-rung Möbius ladder was synthesized by Graf and Lehn (Graf & Lehn; 1975). However, two rungs are not enough to make the graph of a Möbius ladder nonplanar, as we can see from the deformation in Figure 1.7. To get from the first embedded graph to the second, we grab the lower string in the back and pull it down, opening up the boundary of the ladder into a figure eight. Then, we untwist the upper loop and straighten out the picture so that it looks like the third picture in Figure 1.7. Because the graph of the two-rung Möbius ladder can be deformed into a plane, this molecule is not considered to be topologically complex.

Finally, in 1989, a molecular knot was synthesized for the first time by Dietrich-Buchecker and Sauvage (Dietrich-Buchecker & Sauvage, 1989). This was a great achievement, as many chemists had been actively trying to synthesize a knot for over thirty years. Dietrich-Buchecker and Sauvage's molecular knot is made out of 124 atoms and has the form of a *trefoil knot*. We illustrate their molecular knot in Figure 1.8. From a topological point of view, a knot is any embedding of a circle that cannot be deformed into a plane. Although people may say they have knots in their shoelaces, such a knot can be untied by a deformation, and so topologically speaking it is not considered to be a knot. A trefoil knot is the simplest type of knot because it is the only knot with just three crossings, and there cannot exist a knot with fewer than three crossings. The molecular trefoil knot is much smaller than a knotted DNA molecule, so it is

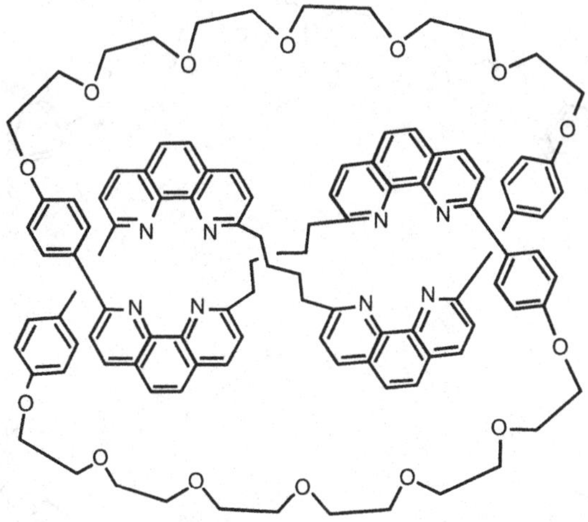

**Figure 1.8.** The molecular trefoil knot

most likely the smallest knot that has ever physically existed. The line segments inside of the hexagons indicate a specific type of bonding. On a chemical level, such hexagons represent benzene rings. The top and bottom arcs of the knot are molecular chains made of oxygens and carbons.

Like the Hopf link, this molecular graph has the property that it cannot be deformed into a plane; however, as an abstract graph it is planar because a different embedding of this graph looks like a planar circle with hexagons attached. This planar embedding can be seen in Figure 1.9.

The molecules that we have listed above are not the only topologically complex molecules that have been synthesized in the past four decades. Rather, we have selected these examples in order to introduce the reader to the history and variety of structures that have been synthesized. Liang and Mislow (1994a) have found many other molecules whose graphs contain $K_{3,3}$, $K_5$, or both. In addition, Liang and Mislow (1994b, 1995a) have found that many proteins naturally contain the nonplanar graphs $K_{3,3}$ or $K_5$ graphs, as well as knots or links.

## Stereoisomers

Walba (1983) coined the term *stereochemical topology* to refer to the synthesis, characterization, and analysis of molecular structures that are topologically nontrivial. Two stereochemical questions that interest topologists are recognizing when one embedding of a graph cannot be deformed to another embedding of the graph and evaluating those properties of embedded graphs that are preserved by deformations.

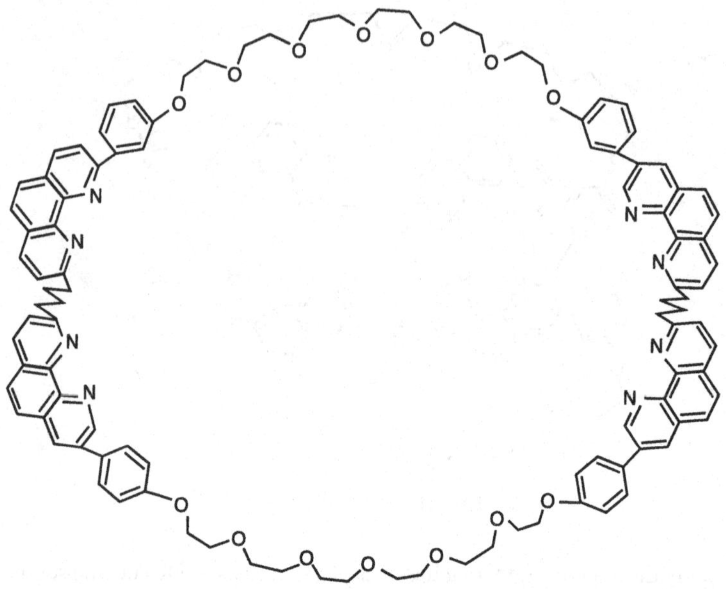

**Figure 1.9.** An unknotted embedding of the graph of the trefoil knot molecule

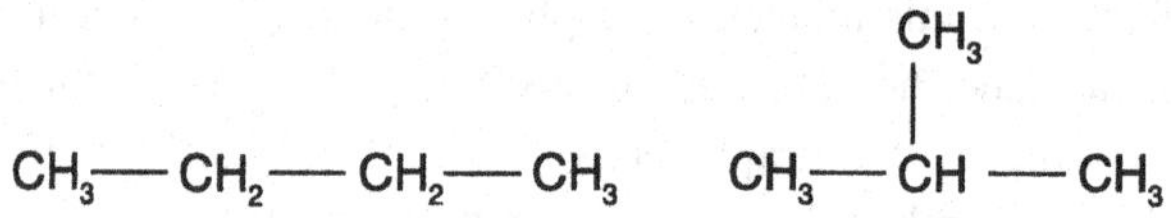

$$CH_3 - CH_2 - CH_2 - CH_3 \qquad CH_3 - \underset{\underset{\displaystyle CH_3}{|}}{CH} - CH_3$$

**Figure 1.10.** A pair of structural isomers

For a given molecule, chemists are interested in enumerating other molecules that are structurally related to it. Such molecules are called the *isomers* of the original molecule. From a chemical, geometric, and topological point of view, we can define three classes of isomers.

1. The *structural isomers* of a given molecule are those molecules that have the same molecular formula but are represented by different molecular bond graphs. In particular, a molecular formula provides a list of all of the atoms contained in a molecule, but it does not specify in what arrangement these atoms are bonded together.

In Figure 1.10, we illustrate a pair of structural isomers called *butane* and *isobutane*, which have different graphs but which both have the molecular formula $C_4H_{10}$.

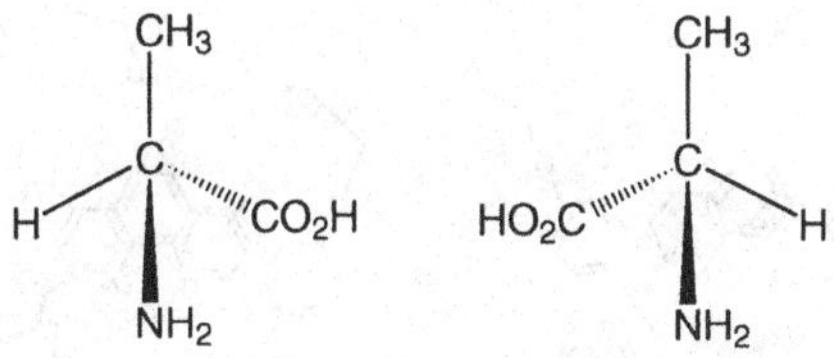

**Figure 1.11.** A pair of rigid stereoisomers

2. The *rigid stereoisomers* of a given molecule are those molecules that have the same abstract graph as the given molecule, but as rigid objects, one could not be picked up and superimposed on the other.

Figure 1.11 illustrates an L-alanine molecule on the left and a D-alanine molecule on the right, which are rigid stereoisomers. These two molecules are mirror images of each other, and the L and the D are there in order to distinguish the two mirror images. The dark triangular segments in Figure 1.11 indicate those edges that are coming out of the plane of the paper toward the reader, the dashed segments indicate those edges that go back behind the plane of the paper, and the ordinary line segments indicate edges that lie in the plane of the paper. In three-dimensional space, the vertices of these graphs lie at the corners of a regular tetrahedron. The reader should observe that if these graphs were flexible, we could deform one to the other, however, because they are rigid graphs, one cannot be picked up and put in the place of the other. For example, if we turn over the graph on the left to try to obtain the graph on the right, then the $CH_3$ and the H will be in the right place but the $NH_2$ will now be behind the page and the $HO_2C$ will be in front of the page.

3. The *topological stereoisomers* of a given molecule are those molecules that have the same abstract graph as the given molecule, but as embedded graphs one cannot be deformed to the other.

Figure 1.12 illustrates a right-handed trefoil molecule (on the right) and a left-handed trefoil molecule (on the left). If you move along the right-handed trefoil knot with your right hand, then you can follow the twists, with your right thumb representing the overcrossings and your right fingers representing the undercrossings. In contrast, if you move along the left-handed trefoil knot with your left hand, then you can follow the twists, with your left thumb representing the overcrossings and your left fingers representing the undercrossings. In 1914, Max Dehn proved that a trefoil knot cannot be deformed to its mirror image (Dehn, 1914). We can use this result to prove that the graphs of the right-handed trefoil molecule and the left-handed trefoil molecule are topological

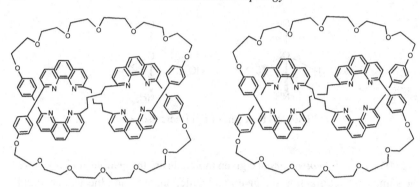

**Figure 1.12.** A pair of topological stereoisomers

stereoisomers. Although this may seem to follow immediately from Dehn's result, to prove it in complete detail requires more care because the molecular graph contains some hexagons rather than just single edges. In Chapter 2, we will discuss how to rigorously prove that the two trefoil molecular graphs are topological stereoisomers.

Note that the topological stereoisomers of a molecule are a subset of the rigid stereoisomers of the molecule, because a graph that cannot be deformed to another graph certainly cannot be rigidly superimposed on the other graph. Most chemists do not distinguish between rigid and topological stereoisomers. Rather, they define a pair of molecules to be *stereoisomers* if they have the same abstract graph but one cannot chemically change itself into the other. Whether this corresponds to our notion of rigid stereoisomers or to our notion of topological stereoisomers depends on to what extent particular bonds can rotate or bend.

Recognizing when two embeddings of a graph are topological stereoisomers is an extension of the well-known problem in knot theory of recognizing when two knots are distinct. A *knot* is an embedding of a circle in three-dimensional space, and two knots are said to be *distinct* if one cannot be deformed to the other. Similarly, two embeddings of the same graph are said to be *distinct* if one cannot be deformed to the other. For example, the two trefoil knotted molecules illustrated in Figure 1.12 are distinct. If we add a finite number of vertices to any knot, we create a graph. Thus the problem of recognizing when two knots are distinct is actually a special case of the problem of recognizing when a pair of embedded graphs are topological stereoisomers. The classification of knots is a very hard problem, so the classification of embeddings of graphs should be at least as hard a problem in topology. In contrast, the vertices of a graph do provide some extra structure, which can help. Furthermore, molecular graphs

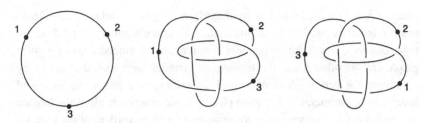

**Figure 1.13.** Different types of topological stereoisomers

often contain labels indicating the type of atom represented by each vertex, and this can sometimes make the problem more tractable.

Two embeddings of a labeled graph may be topological stereoisomers, either as a result of their embeddings as unlabeled graphs or as a result of the particular labeling of the vertices. The three graphs in Figure 1.13 illustrate the two different ways that labeled graphs can be topological stereoisomers. The reason that the first two embedded graphs are topological stereoisomers is because one is knotted and the other one is not. An unknotted circle can never be deformed to a knotted circle. The second embedded graph in Figure 1.13 is the knot that is denoted by the symbol $8_{17}$. Unlike the trefoil knot, most knots do not have names. Rather, knots and links are referred to by numbers. This knot is called $8_{17}$ because eight is the minimum number of crossings with which it can be drawn, and this knot is the seventeenth knot with eight crossings that is listed in the standard knot tables (see, e.g., the tables in Rolfsen, 1976). If we ignore the labeling, the second and third embedded graphs in Figure 1.13 are identical. We shall see as follows that the labeling of the vertices makes this pair of embedded graphs topological stereoisomers. An *orientation* on a knot is a direction associated with the knot, just as the arrow on a one-way street sign provides a direction associated with a road. Traversing the knots in Figure 1.13 by going from vertex 1 to vertex 2 to vertex 3 is one way to specify an orientation on the knots. If we orient the unknotted circle in the clockwise direction, then we can reverse the orientation by turning the circle over. Kawauchi (1979) proved that there is no deformation of the knot $8_{17}$ that reverses an orientation on the knot. Any deformation of $8_{17}$ that exchanged vertices 1 and 3 while holding vertex 2 fixed would reverse the orientation of the knot. Thus by Kawauchi's result, there is no deformation of this knot that exchanges vertices 1 and 3 while holding vertex 2 fixed. So, as labeled embedded graphs, these knots are topological stereoisomers, whereas as unlabeled embedded graphs they are identical.

Any graph that contains a closed loop will have infinitely many topologically distinct embeddings. In fact, for any given embedding of such a graph, we can obtain infinitely many related embeddings by adding more and more little knots in any edge of a closed loop. On one hand, as there are infinitely many different types of knots, it will not be possible to enumerate all embeddings of a given graph. On the other hand, the topological stereoisomers that are created by permuting the vertex labels of a fixed embedding of a graph are finite and hence can be enumerated. For example, we could enumerate all of the labeled embedded graphs that we obtain by permuting the three vertices of the knot $8_{17}$ in Figure 1.13, and then we could determine which of these could be obtained from one another by a deformation of the embedded graph. In this way, we would know all of the topological stereoisomers that are created by permuting the vertex labels of the knot $8_{17}$ in Figure 1.13.

Two vertices of a graph are said to be adjacent if there is an edge connecting them. An *automorphism* (also known as a *self-isomorphism*) of a graph is defined as a bijection (i.e., a function that is one-to-one and onto), taking vertices to vertices in such a way that adjacent vertices are taken to adjacent vertices. We define two labelings of an embedded graph to be *abstractly isomorphic* if there is an automorphism of the graph that takes one labeling to another. If two labelings of an embedded graph are abstractly isomorphic, then they both represent embeddings of the same abstract labeled graph. If we can find an automorphism of a labeled graph that is not induced by a deformation of an embedding of the graph in space, then performing that automorphism on the vertex labels of the embedded graph will create a topological stereoisomer of the original labeled embedded graph. This is how we created the pair of knotted topological stereoisomers in Figure 1.10.

In general, if we wish to enumerate the topological stereoisomers of an embedded labeled graph, then it will help us if we know which graph automorphisms cannot be induced by some deformation of the embedding of a graph. Because there are only finitely many automorphisms of a given labeled graph, determining which labelings of an embedding give us topological stereoisomers should be possible, though not always easy. In particular, there is no known algorithm to determine which automorphisms of a given labeled embedded graph can be induced by a deformation of the graph in three-dimensional space.

In most cases, whether or not a given automorphism can be induced by a deformation depends on the embedding of the graph in three-dimensional space. For example, the automorphism of the unknotted circle in Figure 1.13 that interchanges vertices 1 and 3 while vertex 2 is fixed can be induced by a rotation about a line in the plane bisecting the graph and going through vertex 2. We can denote this automorphism by the pair (13). In general, $(v_{11}, \ldots, v_{1n_1})(v_{21}, \ldots, v_{2n_2}) \cdots (v_{m1}, \ldots, v_{mn_m})$ represents an automorphism that sends each vertex $v_{ij}$ to $v_{ij+1}$ and so on, where the subscripts of each $v_{ij}$

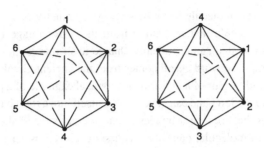

**Figure 1.14.** Topological stereoisomers induced by the automorphism (1234)

are considered mod $n_i$, and all vertices that are not listed are fixed. It follows from our above argument that the automorphism (13) cannot be induced by any deformation of the knot $8_{17}$. In contrast with this example, which depends on the embedding of the graph, there do exist certain graphs with automorphisms that cannot be induced by a deformation of the graph, no matter how the graph is embedded in space. Thus, given any embedding of the graph, performing one of these automorphisms will always yield a topological stereoisomer. The following theorem (Flapan, 1989), proven in Chapter 5, provides an example of such a graph automorphism.

**Theorem 1.1.** *Let* $K_6$ *be the complete graph on six vertices with vertices labeled* 1, 2, 3, 4, 5, *and* 6. *It is not possible to embed* $K_6$ *in three-dimensional space in such a way that the automorphism* (1234) *can be induced by a deformation of the graph in space.*

It follows from this theorem that, no matter how we embed $K_6$ in space, if the vertices are labeled with numerals 1 through 6, then performing the automorphism (1234), which cyclically permutes the first four vertices and fixes the remaining two, will create a topological stereoisomer. For example, the two labeled embedded graphs in Figure 1.14 are topological stereoisomers.

### Chirality

One noteworthy type of stereoisomer occurs as a pair of embedded graphs that are mirror images of one another; see, for example, Figures 1.11 and 1.12. Topologists have developed techniques to help determine when a knot or link is distinct from its mirror image, and some of these techniques have been extended to graphs embedded in space. We will present many of these techniques for knots, links, and embedded graphs in Chapter 2.

Whether or not a molecule has mirror-image symmetry is quite important chemically. A molecule that is distinct from its mirror image is said to be *chiral*, whereas one that can chemically change itself into its mirror image is said to be *achiral*. The word *chiral* comes from the ancient Greek word *cheir*, which means hand. A pair of hands is a prototypical example of a pair of chiral structures, because a left hand can never change itself into a right hand. A pair of chiral molecules that are mirror images of one another are called *enantiomers*.

The concept of molecular chirality was first developed by Pasteur, who observed, in 1848, that mirror-image crystals of tartaric acid rotated polarized light in different directions (Pasteur, 1848). Unpolarized light travels in waves that are not all in the same plane. After a beam of light passes through a polarizing filter, all of the light waves in the beam lie in parallel planes. When the polarized light then passes through a solution containing only one of the enantiomers of a chiral molecule, the planes containing the light waves either will all be rotated to the right or will all be rotated to the left. If one enantiomer always rotates polarized light to the right, then under the same experimental conditions, the other enantiomer will always rotate the light to the left. For example, the molecules L-alanine and D-alanine, illustrated in Figure 1.11, are a pair of enantiomers. Achiral molecules, as well as mixtures of the two enantiomers of a chiral molecule, are both optically inactive; that is, there are no experimental conditions under which they will rotate polarized light. In general, two enantiomers will interact similarly with an achiral molecule but differently with a chiral molecule, just as your right and left foot interact similarly with the ground but differently with a left shoe. A pair of enantiomers often have different properties from one another. For instance, one enantiomer of the molecule called *carvone* smells like caraway, whereas its mirror image smells like spearmint.

Most organisms have a preferred handedness. For example, most helical shells spiral in a right-handed screw. Some types of climbing plants always spiral to the right whereas other types of climbing plants always spiral to the left. Almost all amino acids of proteins in living organisms are chiral. This means that the human body generally reacts differently to one enantiomer of a chiral substance than to the other. It is possible that if proteins, which have the opposite handedness as what they usually have could be synthesized, they could be used to create new pharmaceuticals.

*Nonbiological synthesis* refers to the production of compounds without making use of any living systems. This type of synthesis generally produces a 50:50 mixture of the two enantiomers of a chiral molecule. Such a mixture is called a *racemic mixture*. Often, only one enantiomer of a medicine has the desired effect, and the other passes harmlessly through the body. However, this was not true for the drug Thalidomide, which was given as a racemic mixture to pregnant women in the 1960s to treat morning sickness. The left-handed enantiomer cured the morning sickness while the right-handed enantiomer caused horrible birth defects. In fact, many pharmaceutical products that are currently being produced and sold in a racemic mixture contain one enantiomer that is more effective and the other that causes more undesirable side effects (Thall, 1996). Fortunately, such side effects are not as disastrous as they were for Thalidomide. The reason that most drugs are produced as racemic mixtures is that it is significantly more expensive to produce a compound containing only a single type of enantiomer (called an *optically pure* product). Nonetheless, because of the potential dangers of racemic mixtures, many pharmaceutical companies are beginning to produce optically pure versions of certain drugs. In fact, six of the current top-selling fifteen medications are sold in optically pure form, and some pharmaceutical companies are predicting that in the future the FDA will require all pharmaceutical products to be optically pure (Laird, 1989).

## Chirality of Möbius Ladders

When chemists synthesize a new molecule, they try to give experimental evidence to show that the structure of the molecule is what they claim it to be. In certain cases, providing experimental evidence that a molecule is chiral can help to establish its structure. For example, in order to synthesize a molecular Möbius ladder, Walba, Richards, and Haltiwanger created a molecular ladder and then made the ends of the ladder join together. These ladders did not all join themselves in the same way. Some of the ladders joined together as cylinders, some joined as Möbius ladders with a left-handed half-twist, and some joined as Möbius ladders with a right-handed half-twist.

Researchers at Walba's laboratory wanted to provide experimental evidence that at least some of the ladders joined together with a left-handed or right-handed half-twist in the form of a Möbius strip. In order to do this, they used a technique known as NMR (nuclear magnetic resonance) to show experimentally that some of the synthesized molecules were chiral. As can be seen from Figure 1.15, a cylindrical ladder is its own mirror image and hence is achiral. Walba believed that any structure that has the form of the Möbius ladder in Figure 1.5 must be chiral; however, the Möbius ladders were large enough to have some flexibility, so it was difficult to determine whether or not a Möbius ladder might be able to convert itself to its mirror image by means of some strange deformation. If the embedded graph of a Möbius ladder could be shown to be

**Figure 1.15.** A cylindrical ladder is topologically achiral

topologically distinct from its mirror image, then the experimental evidence of chirality would be consistent with the assumption that some of the molecules were Möbius ladders. Walba conjectured that even with complete flexibility, it was impossible to deform the graph of a molecular Möbius ladder to its mirror image; however, he could not prove this.

Stimulated by Walba's conjecture, Jon Simon (1986), a topologist, proved the following theorem.

**Theorem 1.2.** *The embedded graph of a molecular Möbius ladder with three or more rungs cannot be deformed to its mirror image in such a way that rungs go to rungs and sides go to sides.*

The requirement that rungs go to rungs and sides go to sides is based on the chemical distinction between the carbon–carbon double bonds represented by rungs and the polyether chain represented by the sides of the ladder. At room temperature, when a flexible molecule deforms itself the molecular bonds cannot pass through each other. By Simon's Theorem, the graph of the molecular Möbius ladder cannot be deformed to its mirror image, so a molecular Möbius ladder certainly cannot chemically convert itself to its mirror image. Hence the molecular Möbius ladders are chiral, as Walba suspected. In Chapter 3, we will explain the details of Simon's proof. This theorem inspired topologists to develop tools to detect when an arbitrary embedded graph can be deformed to its mirror image. An embedded graph that cannot be deformed to its mirror image is said to be *topologically chiral*. Simon's results also led to the question of whether the topological chirality of the molecular Möbius ladder is due primarily to the abstract structure of the graph or primarily to the specific embedding of the graph in the form of a Möbius strip, or both. Simon's proof uses both of these properties. The answer to this question turns out to depend on whether the number of rungs of the ladder is odd or even. In Chapter 3, we generalize Simon's techniques in order to prove the following theorem (Flapan, 1989).

**Theorem 1.3.** *No embedding of a Möbius ladder with an odd number of rungs greater than one can be deformed to its mirror image in such a way that rungs go to rungs and sides go to sides.*

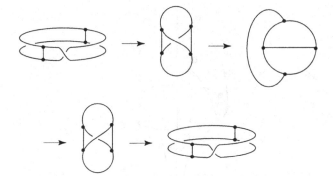

**Figure 1.16.** A deformation of a two-rung Möbius ladder to its mirror image

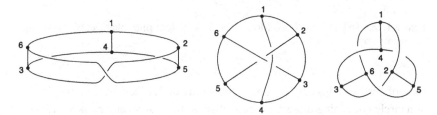

**Figure 1.17.** Three different embeddings of $M_3$

Recall from the beginning of the chapter that the graph of a molecular Möbius ladder with two rungs can be deformed into a plane. Figure 1.16 illustrates how we can combine the deformation of Figure 1.7 with another deformation to take the two-rung Möbius ladder to its mirror image. In particular, after deforming the graph to a plane, we can twist it the other way to get the mirror image.

In contrast with Theorem 1.3, it turns out that if a Möbius ladder has an even number of rungs, then there always exists an embedding of it that can be rotated to its mirror image. In order to understand the different ways that a Möbius ladder can be embedded in space, we first have to understand the abstract graph of a Möbius ladder, independent of any particular embedding. To do this, we denote the abstract graph of a Möbius ladder with $n$ rungs by $M_n$. The left-hand picture in Figure 1.17 is the usual embedding of a molecular Möbius ladder with three rungs. We can see from the left-hand picture in Figure 1.17 that there is a circle of edges connecting vertex 1 to 2, vertex 2 to 3, and so on up to vertex 6, which is connected back to vertex 1. Any embedding of $M_3$ must have the same

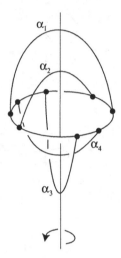

**Figure 1.18.** An embedding of a Möbius ladder with four rungs that can be rotated to its mirror image

pairs of vertices adjacent. Thus any embedding of $M_3$ includes an embedding of a circle containing these six consecutive vertices. Also any embedding has edges connecting vertices 1 to 4, 2 to 5, and 3 to 6. This means that, however $M_3$ is embedded, these pairs of vertices on the circle are also connected by edges. The middle and right-hand pictures in Figure 1.17 illustrate different embeddings of $M_3$. In general, any embedding of $M_n$ contains a circle with $2n$ vertices, as well as additional edges connecting vertices that are halfway around the circle.

In Figure 1.18, we illustrate an embedding of $M_4$ that can be rotated to its mirror image. The horizontal circle in the picture represents the sides of the ladder, and so it contains the eight vertices that are the endpoints of the four rungs labeled $\alpha_1$, $\alpha_2$, $\alpha_3$, and $\alpha_4$. We can rotate this embedding of $M_4$ by 90° about a vertical axis in order to obtain its mirror image reflected through a central horizontal plane. This 90° rotation together with a mirror reflection through the plane takes the horizontal circle to itself, exchanging the rung $\alpha_1$ with the rung $\alpha_3$, and exchanging the rung $\alpha_2$ with the rung $\alpha_4$. By adding symmetric pairs of rungs above and below the plane, we can create a similar embedding of a Möbius ladder with any even number of rungs, which can also be rotated to its mirror image. Thus, for any even $n$, the topological chirality of the embedded graph of the molecular Möbius ladder with $n$ rungs is due to the particular embedding of the graph in the form of a Möbius strip, rather than to the abstract structure of the graph.

### Intrinsic Chirality

A property of a graph is said to be *intrinsic* to the graph if it does not depend on any particular embedding of the graph in three-dimensional space. For example, the nonexistence of a planar embedding of a graph is an intrinsic property. There are also other intrinsic properties of a graph that are of topological interest. We will discuss some of these in Chapter 5. We are particularly interested in when the property of chirality is intrinsic to a graph. Any graph that has the property that no embedding of it can be deformed to its mirror image is said to be *intrinsically chiral.*

Note that, for simplicity, when we draw a Möbius ladder we never include the vertices of valence 2 which were included in Figure 1.5. However, we imagine that the edges making up the sides of the ladder are colored with a different color from the edges that make up the rungs of the ladder. Theorem 1.3 implies that any colored Möbius ladder with an odd number of rungs greater than one is intrinsically chiral. Chemically, we could define a molecule to be intrinsically chiral if it, and all of its topological stereoisomers (both real and hypothetical), are chiral. Thus knowing that a molecular graph is intrinsically chiral means that we do not have to evaluate the chirality of each individual stereoisomer. In Chapters 3 through 6, we shall prove that a number of graphs are intrinsically chiral. For example, in Chapter 5, we shall prove that for any natural number $p$, the complete graph on $4p + 3$ vertices is intrinsically chiral. In Chapters 3 and 6, we also prove that several molecular graphs are intrinsically chiral in addition to the Möbius ladders with an odd number of rungs.

Intrinsic chirality was characterized by Flapan and Weaver (1996) for a large class of graphs in terms of whether or not the graphs have certain types of automorphisms. However, these automorphisms can be hard to recognize, and there is no known algorithm for determining whether or not a particular graph is intrinsically chiral.

### Euclidean Rubber Gloves

It would be convenient to be able to determine from a complex molecular graph whether or not the molecule it represents is chemically chiral. For this reason it would be nice to have a definition of chirality that had the property that a molecule would be chemically chiral if its molecular bond graph were chiral according to this definition. Because chemists are used to working with rigid molecules, the following definition is often used by chemists.

**Definition.** A molecular bond graph is *achiral* if it can be rigidly rotated to its mirror image. Otherwise it is *chiral.*

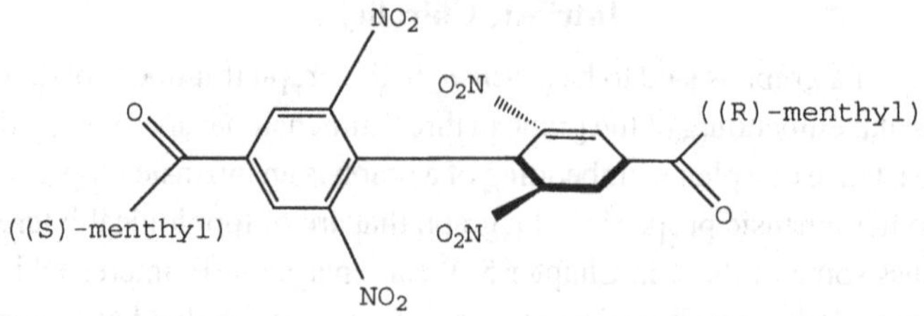

**Figure 1.19.** A molecule that can change itself into its mirror image but cannot be rotated to its mirror image

This definition is equivalent to saying that a molecular bond graph is achiral if it can be superimposed on its mirror image. If a molecular bond graph is achiral according to this definition, then as a rigid object it is identical to its mirror image. In particular, this means that on a chemical level the molecule can change itself into its mirror image, so it is chemically achiral. In contrast, if a molecular bond graph is chiral according to this definition, it does not necessarily follow that the molecule is chiral. That is, just because a molecular graph cannot be rigidly superimposed on its mirror image does not mean that the molecule it represents cannot change itself into its mirror image. If the molecule is long enough that it is flexible, it might be able to deform itself to its mirror image even if no rotation will take it to its mirror image.

Even apart from such flexible molecules, some relatively rigid molecules can change themselves into their mirror images if they can twist around a specific bond or bonds. In 1954, Mislow exhibited the first such molecule that could chemically change itself into its mirror image but whose molecular bond graph was chiral according to this definition (Mislow & Bolstad, 1955). Mislow's molecule, which is illustrated in Figure 1.19, is a derivative of biphenyl. The two menthyls on either end of the graph are mirror images of each other. In particular, the S and the R stand for *sinister* and *rectus*, which are the latin words for left and right, respectively. It is important to note that these two letters do not correspond to the direction in which polarized light is rotated; rather, they correspond to the geometric structure of the two forms of the menthyl. The hexagon on the left in the figure, is vertical while the hexagon on the right is horizontal, with the dark edges sticking out in front of the plane of the paper and the dashed edges going behind the plane of the paper. This molecular graph has vertical propellers on each end that rotate simultaneously by 90°. When the molecule rotates about a horizontal axis by 90°, and the propellers simultaneously rotate by another 90°, we obtain the mirror image of the molecule through a vertical mirror cutting across the central bond of the molecule. Observe that in our figure the $NO_2$ and the $O_2N$ represent the same molecular subunit. However, we write each one with the N closest to the bond with the benzene ring in order to indicate in both cases that the benzene is bonded with the nitrogen atom, rather than with either of the two oxygens. Apart from the simultaneous rotation of the propellers, the molecule is rigid, so it can never assume a position that can simply be rotated to its mirror image. Observe that if the central bond were not rigid, then the molecular graph could become planar and then it would be its own mirror image by reflecting through the plane that contains the graph.

We refer to a structure that is in a position that can be rotated to its mirror image or is already its own mirror image as a *symmetry presentation* of that structure. Thus Mislow's biphenyl derivative can change itself into its mirror image, but it is chemically unable to take the form of a symmetry presentation.

Walba (1983) has referred to any molecule that can change itself into its mirror image but has no chemically accessible symmetry presentation as a *Euclidean rubber glove*. The idea behind this terminology originated with an unpublished paper of Van Gulick written in 1960 and eventually published in 1993. Van Gulick (1993), observed that a right-handed rubber glove can be turned inside out to become a left-handed rubber glove, but at no point during the turning inside out process can the rubber glove be superimposed on its mirror image. If the glove were to be stretched and flattened out so that it lay in a plane, then it would be its own mirror image. However, the glove does not become flattened out while it is being turned inside out. Similarly, if we allow the graph of Mislow's molecule to become flexible, it could also be flattened into a plane, but on a molecular level this does not occur. Walba added the word "Euclidean" to the idea of a rubber glove in order to imply that, as with Mislow's molecule, apart from some bonds that can rotate, the molecule actually must remain rigid as a geometric object does in Euclidean geometry.

## Topological Rubber Gloves

Extending the notion of Euclidean rubber gloves to more flexible structures, Walba asked whether a flexible embedded graph could exist that could be deformed to its mirror image but could not be deformed to a symmetry presentation. He called such a structure a *topological rubber glove*, as the flexible analog of a Euclidean rubber glove. To see the analogy, observe that a Euclidean rubber glove is a molecule that can chemically change itself into its mirror image but has no chemically accessible symmetry presentation, whereas a topological rubber glove is an embedded graph (which may or may not be a molecular bond graph) that can be deformed to its mirror image but has no topologically

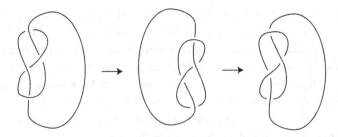

**Figure 1.20.** A deformation of a figure eight knot to its mirror image

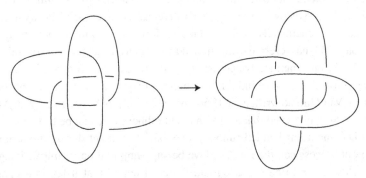

**Figure 1.21.** A presentation of the figure eight knot that can be rotated to its mirror image

accessible symmetry presentation. The embedded graph of Mislow's Euclidean rubber glove is not a topological rubber glove, because if the graph were flexible then it could be deformed to a planar symmetry presentation.

It is sometimes hard to determine whether or not a flexible structure has a symmetry presentation. Walba initially conjectured that the figure eight knot was a topological rubber glove because he could see how to deform it to its mirror image, but he could not find a symmetry presentation for it. Figure 1.20 illustrates a deformation of the figure eight knot to its mirror image. To get from the first picture to the second picture, you rotate the knot by 180° about a vertical axis. Then to get to the third picture, you flip the long string over the knotted arc without moving the knotted arc. This final picture is the mirror image of the first picture, where the mirror is in the plane of the paper. This means that the mirror has the effect of interchanging all of the overcrossings and undercrossings.

The figure eight knot is not a topological rubber glove because it can be deformed to a position that can be rotated to its mirror image. A symmetry presentation for the figure eight knot is illustrated in Figure 1.21. (The reader

**Figure 1.22.** Knot $8_{17}$ is a topological rubber glove

should check that the knots in Figures 1.20 and 1.21 are the actually the same.) We obtain the mirror image of this presentation by rotating the picture by 90° about an axis perpendicular to the page that goes through the center of the picture. Here the mirror is again in the plane of the paper, so the overcrossings and undercrossings in the first and second pictures are interchanged. Notice that this symmetry presentation of the figure eight knot has more crossings than the original picture of the figure eight knot. It is often the case that a symmetry presentation of a knot has more than the minimum number of crossings for that knot. There is no known algorithm for finding a symmetry presentation for any knot which has one, or even for determining whether an arbitrary knot has a symmetry presentation.

Figure 1.22 illustrates a deformation of the knot $8_{17}$ to its mirror image, which is similar to the deformation of the figure eight knot in Figure 1.20. Here the knot $8_{17}$ is drawn somewhat differently than it was in Figure 1.13; however, it is possible to deform the projection of Figure 1.13 to obtain that of Figure 1.22. To get from the first picture in Figure 1.22 to the second picture, we rotate the knot by 180° about an axis perpendicular to the page going through the center of the knot. Then to get from the second picture to the third picture, we pull the lower string over the top of the knot. Again the third picture is the mirror reflection of the first picture, where the mirror is in the plane of the paper. In Chapter 4, we prove that the knot $8_{17}$ has no symmetry presentation and hence is a topological rubber glove. This was the first example of a topological rubber glove and, in fact, turns out to be the simplest knot that is a topological rubber glove.

Starting with a Möbius ladder that is embedded in a symmetry presentation, we can create a graph that is a topological rubber glove, and actually has the stronger property that no embedding of the graph can have a symmetry presentation. Such a graph is illustrated in Figure 1.23. Notice that the left-hand side of the graph is actually the topologically achiral embedding of the Möbius ladder with four rungs that we saw in Figure 1.18. The right-hand side of the graph is a triangle that is its own mirror image. We can deform the graph in Figure 1.23 to its mirror image by rotating the left-hand side by 90° about a vertical axis (as we did in Figure 1.18), while keeping the right-hand side fixed. Here, the mirror is a horizontal plane that contains the horizontal circle and the

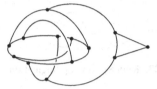

**Figure 1.23.** This embedded graph can be deformed to its mirror image, but no embedding of it can be rotated to its mirror image

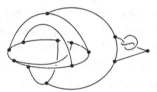

**Figure 1.24.** A chiral embedding of the graph from Figure 1.23

vertex of the triangle. In Chapter 4, we will present a proof that no embedding of this graph can be rotated to its mirror image.

We could say that this graph is an intrinsic topological rubber glove, because it has an embedding that is topologically achiral but has no embedding that has a symmetry presentation. However, this does not mean that every embedding of the graph is a topological rubber glove, because not every embedding of this graph can be deformed to its mirror image. In particular, we could embed the graph with a little chiral knot in one of the edges, as illustrated in Figure 1.24. Because the knot is chiral, this embedding cannot be deformed to its mirror image. Only a graph containing no closed circuits could have the property that every embedding of it can be deformed to its mirror image. However, a graph with no closed circuits is a tree or a collection of disjoint trees, and every embedding of such a graph can be deformed to a planar symmetry presentation. Hence there cannot be a graph such that every embedding of it is a topological rubber glove.

Although the knot and the graph in Figures 1.22 and 1.23 illustrate that topological rubber gloves can exist as topological structures, neither of these examples is a molecular graph. However, in 1992, Du and Seeman synthesized a single-stranded DNA molecule in the form of a figure eight knot (Du & Seeman, 1992), which can be shown to be a topological rubber glove. Furthermore, in 1997, Chambron, Sauvage, and Mislow designed and synthesized a molecular link that was not only a topological rubber glove but was chemically achiral (Chambron et al., 1997). We illustrate their molecule in Figure 1.25. We shall

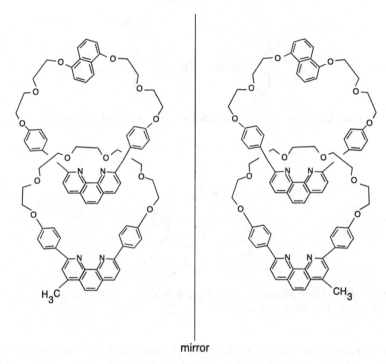

**Figure 1.25.** A topological rubber glove that is chemically achiral

discuss these molecules and prove in detail that they are a topological rubber gloves in Chapter 4.

Though various examples are known, no techniques have yet been developed to determine whether or not an arbitrary embedded graph is a topological rubber glove.

## Nonrigid Symmetries

By analogy with Euclidean and topological rubber gloves, we could ask if there are rigid stereoisomers that are not topological stereoisomers; that is, one embedded graph that can be deformed to another but the first one cannot be rotated to the second one. The pair of graphs presented in Figure 1.11 (which is redrawn in Figure 1.26) is a simple example of a pair of rigid stereoisomers that are not topological stereoisomers. These graphs are abstractly isomorphic and one can be deformed to the other, but one cannot be rigidly rotated to the other.

We can state the above assertion in terms of automorphisms by saying that there is an automorphism of the vertices of the first graph that can be induced by

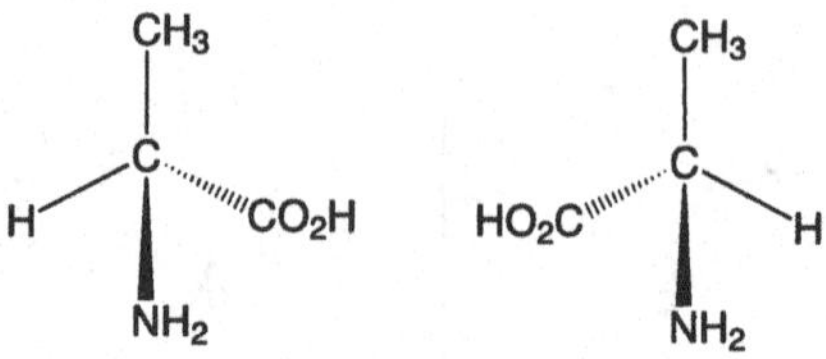

**Figure 1.26.** Rigid stereoisomers that are not topological stereoisomers

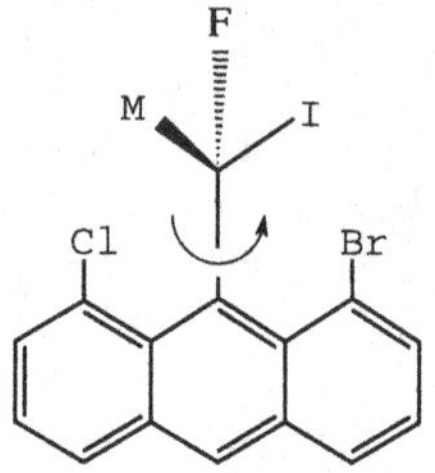

**Figure 1.27.** Automorphism $\varphi$ can be induced by a deformation but not by a rigid motion of any embedding of this graph

a deformation of this embedding, but cannot be induced by a rigid motion of the embedding. One particular automorphism, which has this property, is the one that interchanges vertex H with vertex $NH_2$. If we perform this automorphism on the first graph, then the first graph could be rotated to the second graph. Of course, if the graph were flexible we could deform it to a planar cross with the $CO_2H$ at the bottom and the H and the $NH_2$ on opposite sides, so that the automorphism switching H and $NH_2$ is induced by simply turning the graph over. In particular, if we re-embed the graph as a cross, rather than a tetrahedron, then this automorphism would be induced by a rigid motion. From a chemical point of view, this graph cannot take the form of a planar cross. In particular, chemically, the molecules in Figure 1.26 are distinct and thus are stereoisomers.

By analogy with the example in Figure 1.23, we can consider a molecular graph that has an automorphism that can be induced by a deformation, but, no matter how the graph is embedded in space, that automorphism cannot be induced by a rigid motion of space. For example, consider the molecular graph illustrated in Figure 1.27, with the automorphism $\varphi$, which cyclically permutes the three vertices labeled M, I, and F. For this particular embedding, the automorphism $\varphi$ is induced by the deformation that rotates the top of the graph by 120° while keeping the rest of the graph fixed. In Chapter 6, we will prove that no matter how this graph is embedded in space, there is no rigid motion that will induce this automorphism. Thus, no matter how this graph is embedded in space, performing this automorphism will give us a rigid stereoisomer. However, for the particular embedding illustrated in Figure 1.27, performing this automorphism will not give us a topological stereoisomer because the automorphism can be induced by a deformation. Furthermore, because this partial rotation is chemically achievable, performing this automorphism will not give us a chemical stereoisomer.

The graphs illustrated in Figures 1.23 and 1.27 both have symmetries that can be induced by a deformation, but, no matter how the graph is embedded, this symmetry cannot be induced by a rigid motion. We would like to understand, in general, which graphs have the property – that they have a symmetry that can be induced by a deformation of some embedding of the graph, but, no matter how the graph is embedded, that symmetry cannot be induced by a rigid motion of space. An important property of a rigid motion of a graph in space is that if we perform the motion some finite number of times, then every point in space will return to its original position. For a given rigid motion, the smallest such number is said to be the *order* of the rigid motion. For example, if we perform a rotation of 120° three times, then every point returns to its original position. Such a rotation is therefore said to have order three. In general, a rotation of $360°/n$ has order $n$, and a reflection has order two. In contrast, a deformation does not necessarily have finite order. For example, the deformation that rotates only the top of the graph in Figure 1.27 by 120° is not of finite order. If we perform this deformation three times, then every point on the graph will return to its original position but points in space that are near the bond around which we are rotating will not return to their original positions.

In order to understand what is special about the graphs in Figures 1.23 and 1.27, we have the following definition.

**Definition.** A graph is said to be *n connected* if at least *n* vertices must be removed, together with the edges containing them, in order to disconnect the graph or reduce it to a single vertex.

We are particularly interested in three-connected graphs. Intuitively, a graph that is not three connected may be embedded so as to have pieces that rotate independently of the graph, like beads on a necklace. Most complicated graphs are either three connected or made up of three-connected pieces. For example, all Möbius ladders and complete graphs with at least four vertices are three connected. Even though graphs that are three connected are more complex than graphs that are not, they are better behaved in the sense that they do not permit this type of rotation around one or two vertices. Observe that the graphs in Figure 1.23 and Figure 1.27 are not three connected, because we can remove two or one vertices, respectively, to disconnect them. Also, observe that each has a partial rotation like a bead rotating on a necklace.

Although molecules live in ordinary three-dimensional space, it is sometimes mathematically convenient to add an extra point to space, which we will call the point at *infinity*. Symbolically, we let $\mathbb{R}^3$ denote Euclidean three-dimensional space and we let $S^3$ denote $\mathbb{R}^3$ with a point added at $\infty$. In Chapter 6, we will prove that any three-connected graph embedded in $\mathbb{R}^3$ in such a way that it has an automorphism that can be induced by a deformation also has an embedding in $S^3$ such that this same automorphism can be induced by a rigid motion of $S^3$. In $\mathbb{R}^3$ the situation is somewhat more complicated, and there is no known characterization of those graphs that have automorphisms that can be induced only by deformations and not by rigid motions.

The proof of the result described above is rather complicated and uses some deep results in three-dimensional topology. However, it has the following corollary, which is elementary to state and not hard to prove. The proof will also be given in Chapter 6.

**Corollary 1.4.** *Any nonplanar graph that has no order-two automorphism which is label-preserving is intrinsically chiral.*

Just as we defined the order of a rigid motion, we can define the *order* of an automorphism of a graph as the smallest number such that performing the automorphism that many times returns every vertex to its original position. An automorphism is said to be *label-preserving* if it takes any labeled vertices to vertices with the same labels. In particular, a label-preserving automorphism of a molecular graph takes atoms of a given type to atoms of the same type. Corollary 1.4 enables us to easily show that many graphs are intrinsically chiral. For example, in Chapter 6, we will use this corollary to prove the intrinsic chirality of the Simmons–Paquette molecular graph as well as the graph of the molecule ferrocenophan, illustrated on the left and right, respectively, of Figure 1.28.

In addition to this corollary, the theorem in Chapter 6 allows us to characterize which automorphisms of a complete graph can be induced by a deformation of some embedding of the graph in $\mathbb{R}^3$. It follows from this characterization theorem that many automorphisms of a complete graph with at least seven vertices cannot be induced by any deformation, no matter how the graph is embedded. We can use these automorphisms to obtain many topological stereoisomers similar to those illustrated in Figure 1.14 of this chapter.

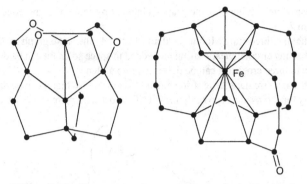

**Figure 1.28.** These molecular graphs are intrinsically chiral

## Exercises

1. Explain why every achiral molecule, which is a topological rubber glove, is also a Euclidean rubber glove, but not the converse.
2. Let $n$ be a natural number. Find a graph that has automorphisms of every order up to $n$.
3. Find a graph that has $n$ different order two automorphisms, and no automorphism of order more than two.
4. Show that the following molecular graph can be deformed into the plane if it is completely flexible.

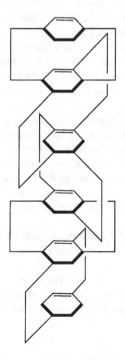

5. Find planar embeddings for the graphs obtained by removing one edge from $K_5$ and from $K_{3,3}$.
6. Prove that the three-rung Möbius ladder is abstractly the same graph as $K_{3,3}$.
7. Prove that any circular elastic string embedded in space so that it can be drawn with fewer than three crossings can be deformed into a plane.
8. Prove that any circular elastic string embedded in space so that it can be drawn with precisely three crossings can be deformed either to a planar circle or to a trefoil knot.

9. Prove that the two trefoil knots with labeled vertices given below are not topological stereoisomers.

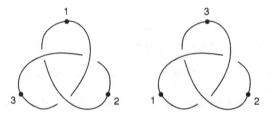

10. Enumerate all of the labeled embedded graphs that we obtain by permuting the three vertices of the knot $8_{17}$ in Figure 1.13, and determine which ones could be obtained by a deformation of the graph.
11. Consider a circle containing four vertices numbered consecutively by numerals 1 through 4. How many labelings of this graph are there by these numbers that are abstractly isomorphic to this one? How many ways are there to label the vertices with numerals 1 through 4 so that no labeling is abstractly isomorphic to any other labeling?
12. Let $K_6$ denote the complete graph with six vertices labeled with numerals 1 through 6. Find an embedding of $K_6$ in three-dimensional space such that the automorphism (123)(456) is induced by a deformation of the graph.
13. Prove that graphs $K_{3,3}$, $K_5$, and $K_6$ are not intrinsically chiral.
14. Show that the knot in Figure 1.20 can be deformed to the knot in Figure 1.21, and that the knot in Figure 1.13 can be deformed to the knot in Figure 1.22.
15. Prove that the abstract graph of the ferrocenophan molecule, which is illustrated in Figure 1.28, is nonplanar.
16. Find embeddings of $K_4$ and $K_5$ such that for each of your embedded graphs, every automorphism of the graph can be induced by some deformation and/or reflection of the embedding.
17. Find an embedding of $K_{3,3}$ such that every automorphism of $K_{3,3}$ can be induced by some deformation and/or reflection of this embedding.
18. Prove that the graphs in Figure 1.28 have no order-two label preserving automorphisms.

19. Prove that for every natural number $n$, the complete graph $K_n$ is $k$ connected if $k < n$, but $K_n$ is not $n$ connected.
20. Prove that for every natural number $n$, the Möbius ladder $M_n$ is three connected.
21. Prove that the molecular graph of Kuratowski cyclophane (which is illustrated below) can be deformed to a symmetry presentation. This molecule was synthesized by Chen et al. (1995) and shown to be chemically achiral.

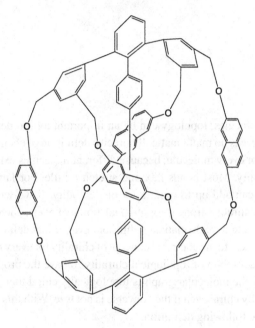

22. Let $G$ denote a circle with $n$ vertices. Prove that the order of any automorphism of $G$ is either two or divides $n$.
23. Prove that if we don't distinguish between rungs and sides then the embedded graph of a 3-rung molecular Mobius ladder can be deformed to its mirror image. Be sure to give a step-by-step deformation with detailed explanations.
24. Start with an embedding of $M_4$ in the form of a Mobius strip. Give a step-by-step deformation of your $M_4$ so that the circle representing the sides of the ladder now lies within the plane.
25. Let $M_3$ be embedded in the form of a Mobius strip. Find automorphisms $f$ and $g$ which are induced by deformations of $M_3$ such that the order of $f$ is 6, the order of $g$ is 2, and $g \circ f \neq f \circ g$ (that is, performing $f$ and then $g$ has a different effect on $M_3$ than performing $g$ and then $f$).
26. Prove that any automorphism $f$ of a graph $G$ preserves the valence of each vertex. That is, if $v$ is a vertex of $G$ which has precisely $n$ edges attached to it, then $f(v)$ also has precisely $n$ edges attached to it. Also prove that if vertices $v$ and $w$ are separated by $n$ vertices in $G$, then the vertices $f(v)$ and $f(w)$ are also separated by $n$ vertices in $G$.

# 2

# Detecting Chirality

For complex molecules, topology can be an important tool to detect chirality. Nonetheless, there is no mathematically precise definition of chemical chirality that will work for every molecule, because different molecules exhibit different levels of flexibility. Most bonds flex and stretch a little. For large molecules these little bits can add up to quite a lot of flexibility. If we wish to discuss the concept of chirality from a topological point of view, then we need to commit ourselves to a mathematical definition, even if that definition does not correspond precisely to the chemical concept of chirality for every molecule. We would like our definition of topological chirality to have the property that, for any molecule, if the molecular graph is topologically chiral then the molecule will be chemically chiral even if the converse is not true. With this goal in mind, we start with the following definition.

**Definition.** A graph embedded in three-dimensional space is *topologically achiral* if it can be deformed to its mirror image. Otherwise it is *topologically chiral*.

If a molecule can convert itself to its mirror image, then the transformation that the molecule goes through to get to the mirror image corresponds to a deformation from the graph to its mirror image. In particular, though individual bonds of the molecule may rotate or flex, bonds do not pass through one another at room temperature. Analogously, we do not permit one edge of a graph to pass through another edge during a deformation of a graph. Hence if there is no deformation taking a molecular graph to its mirror image, then the molecule it represents is necessarily chemically chiral. In contrast, if a molecular graph is topologically achiral then we still do not know whether or not the molecule can actually change itself into its mirror image, because the specific deformation that transformed the graph may not be chemically achievable. So, the concept of

32

topological chirality is chemically useful; however, the concept of topological achirality is not always chemically as useful.

## Defining Chirality

We need to formalize our definition of topological chirality by being more rigorous about what we mean by "deform". To do this we first develop a topological notion of equivalence. For those readers who are not already familiar with the concept of homeomorphic topological spaces, we restrict our discussion to homeomorphic subspaces of Euclidean spaces, which we define below. This will certainly be sufficient for our purposes, as all of our spaces will be contained in some Euclidean space. We begin with the definition of continuity for functions of Euclidean $n$-dimensional space, $\mathbb{R}^n$. If $x$ and $y$ are points in $\mathbb{R}^n$, then the distance from $x$ to $y$ is denoted by $\|x - y\|$.

**Definition.** Let $A$ be a subset of $\mathbb{R}^n$ and let $B$ be a subset of $\mathbb{R}^m$. Let $h : A \to B$ be a function. We say that $h$ is *continuous* if, for every $a \in A$ and for every $\varepsilon > 0$, there is a $\delta > 0$, such that for every $x \in A$ if $\|x - a\| < \delta$ then $\|h(x) - h(a)\| < \varepsilon$.

This definition is probably familiar to the reader, at least for functions from $\mathbb{R}$ to $\mathbb{R}$. Using this concept, we can define what it means for two subsets of Euclidean spaces to be topologically equivalent.

**Definition.** Let $A$ be a subset of $\mathbb{R}^n$ and let $B$ be a subset of $\mathbb{R}^m$. Let $h : A \to B$ be a function. We say that $h$ is a *homeomorphism* if $h$ is continuous, and $h$ has a continuous inverse. In this case, we say that $A$ and $B$ are *homeomorphic*. If there is a homeomorphism $h : \mathbb{R}^n \to \mathbb{R}^n$ such that $h(A) = B$, then we write $h : (\mathbb{R}^n, A) \to (\mathbb{R}^n, B)$.

Those readers with a background in topology should note that, in order not to get too technical, we do not explicitly state that our homeomorphisms are piecewise linear or differentiable. However, readers who are concerned about such things should assume that all of our homeomorphisms are either piecewise linear or diffeomorphisms according to whichever condition is more appropriate in the given context.

Intuitively, two subsets $A$ and $B$ of Euclidean spaces are homeomorphic if creatures living in the spaces, who cannot compare distances, could not tell the difference between the two spaces. For example, a square and a circle of any size are homeomorphic, whereas a circle and a line are not homeomorphic.

If $h : A \rightarrow A$ is a homeomorphism, then we say that $h$ is a homeomorphism of $A$; and if $h : (\mathbb{R}^n, A) \rightarrow (\mathbb{R}^n, A)$ is a homeomorphism, then we say that $h$ is a homeomorphism of $(\mathbb{R}^n, A)$. For example, a rotation $h$ is a type of homeomorphism of $\mathbb{R}^3$ that takes the axis $A$ of rotation to itself, so we write $h : (\mathbb{R}^3, A) \rightarrow (\mathbb{R}^3, A)$.

We are interested in homeomorphisms of $\mathbb{R}^3$ that take a graph to itself. In particular, if $G$ is a graph, we say that $h$ is a homeomorphism of $(\mathbb{R}^3, G)$ and write $h : (\mathbb{R}^3, G) \rightarrow (\mathbb{R}^3, G)$, if $h : \mathbb{R}^3 \rightarrow \mathbb{R}^3$, $h(G) = G$, and $h$ takes vertices to vertices and edges to edges. Note that the hypothesis $h(G) = G$ does not imply that each point of $G$ is necessarily fixed by $h$. The *valence* of a vertex is the number of edges containing that vertex. If all of the vertices of $G$ have a valence of at least three, then requiring that $h(G) = G$ actually forces $h$ to take vertices to vertices and edges to edges. However, many molecular graphs have vertices of valence two, so that is why we have to explicitly require that $h$ takes vertices to vertices and edges to edges.

If $G$ is a graph, then we can consider $G$ as a subset of $\mathbb{R}^3$ so that we can apply our definition of homeomorphism. Formally, an *embedding* of $G$ in $\mathbb{R}^3$ is the image of a homeomorphism $h : G \rightarrow G'$ where $G'$ is a specific subset of $\mathbb{R}^3$. We think of an embedding as a positioning of $G$ in $\mathbb{R}^3$, and we often abuse notation by letting $G$ denote both the abstract graph and the particular embedding of the graph in $\mathbb{R}^3$.

Suppose that $S$ and $T$ are subsets of $\mathbb{R}^n$ and $\mathbb{R}^m$, respectively. We define the *product* set $S \times T = \{(s, t) \in \mathbb{R}^{n+m} | s \in S, t \in T\}$. For example, let $I$ denote the unit interval $[0, 1]$; then $I \times I$ is a unit square in $\mathbb{R}^2$ and $I \times I \times I$ is a unit cube in $\mathbb{R}^3$. A product that is somewhat more difficult to imagine is $S^1 \times D^2$, where $S^1$ represents the unit circle $\{(x, y) | x^2 + y^2 = 1\}$ and $D^2$ represents the unit disk $\{(a, b) | a^2 + b^2 \leqslant 1\}$. According to our definition of the product, $S^1 \times D^2$ should live in $\mathbb{R}^4$; however, $S^1 \times D^2$ is homeomorphic to the solid that we obtain by spinning the disk $D^2$ around a disjoint axis in $\mathbb{R}^3$. This figure looks like a doughnut and is called a *solid torus*. Every point in the solid torus can be expressed uniquely by a pair $(s, t)$, where $s$ is a point in $S^1$ and $t$ is a point in $D^2$. A *torus* is defined as the product $S^1 \times S^1$ and can be visualized as the surface of a doughnut.

Now we give a precise definition of what we mean by a deformation in three-dimensional space.

**Definition.** Let $A$ and $B$ be contained in a set $M$, which is a subset of $\mathbb{R}^n$. We say that $A$ is *ambient isotopic* to $B$ in M if there is a continuous function $F : M \times I \rightarrow M$ such that for each fixed $t \in I$ the function $F(x, t)$ is a homeomorphism, $F(x, 0) = x$ for all $x \in M$, and $F(A \times \{1\}) = B$. The function $F$ is said to be an *ambient isotopy*.

**Figure 2.1.** A planar circle and a trefoil knot are homeomorphic but not ambient isotopic

For this definition, we often think of $t$ as representing time. Thus at time $t = 0$, the ambient isotopy $F$ fixes every point of $M$. Then as time passes, $F$ progressively takes the set $A$ closer and closer to the set $B$. Finally, at time $t = 1$, $F$ takes the set $A$ onto the set $B$. The requirement that at each fixed time $t$ the function $F(x, t)$ is a homeomorphism guarantees that at no time does $F$ make part of $A$ pass through itself. Readers should convince themselves that this definition corresponds to our intuitive conception of a deformation. We need to have this formal definition of ambient isotopy in order to guarantee that our proofs rest on a solid mathematical foundation. However, in practice we generally prove that two knots, links, or embedded graphs are ambient isotopic by drawing a sequence of pictures to illustrate a deformation of one to the other, not by writing formulas representing the ambient isotopy. From now on we shall use the words *deformation* and *ambient isotopy* interchangeably.

It is important to understand the difference between saying that sets $A$ and $B$ are homeomorphic and saying that they are ambient isotopic. If $A$ and $B$ are ambient isotopic in $\mathbb{R}^3$, then they are necessarily homeomorphic, since $F(x, 1)$ is a homeomorphism that takes $A$ onto $B$. In contrast, there can exist homeomorphic subsets of $\mathbb{R}^3$ that are embedded in topologically distinct ways, so that there is no deformation taking one to the other. For example, there is a homeomorphism from a unit circle $S^1 = \{(x, y) \in \mathbb{R}^2 \mid x^2 + y^2 = 1\}$ onto a trefoil knot $K$ (see Figure 2.1). However, $S^1$ and $K$ are not ambient isotopic in $\mathbb{R}^3$ because there is no way to deform a knotted circle to an unknotted circle if we do not permit the knot to cross through itself. Although the knot $K$ is homeomorphic to a planar circle, the complement of $K$ in $\mathbb{R}^3$ is not homeomorphic to the complement of the planar circle. If we make our $K$ and $S^1$ out of rope, we can think of the complement of $K$ and $S^1$ in $\mathbb{R}^3$ as being represented by a space with a tunnel cut out of it. On one hand, an ant going through a long tunnel cannot tell the difference between a tunnel that is knotted and one which is not. This corresponds to the fact that $K$ and $S^1$ are homeomorphic. On the other hand, a bead with a knotted tunnel cut in it is quite different from one with an unknotted tunnel. Because the complements of $S^1$ and $K$ are different, no homeomorphism from $S^1$ to $K$ can be extended to a homeomorphism taking $\mathbb{R}^3$ to itself. In particular, this means that $K$ is not ambient isotopic to $S^1$ in $\mathbb{R}^3$.

Using the concept of ambient isotopy, we can formalize our definition of topological chirality as follows.

**Definition.** A graph $G$ embedded in $\mathbb{R}^3$ is *topologically achiral* if $G$ is ambient isotopic to its mirror image; otherwise $G$ is *topologically chiral.*

In many of the theorems that we will prove, it turns out to be more convenient to work in the three-dimensional sphere $S^3$ rather than in $\mathbb{R}^3$ because there is more topological machinery available in the three-dimensional sphere. To understand $S^3$, we first recall that the unit circle, $S^1$, is the set of all points in $\mathbb{R}^2$ that are one unit from the origin, and the two-dimensional sphere, $S^2$, is the set of all points in $\mathbb{R}^3$ that are one unit from the origin. Analogously, we can define the *three-dimensional sphere*, $S^3$, as the set of all points in $\mathbb{R}^4$ that are a distance of one unit from the origin. For those with some background in topology, an equivalent definition is that the three-dimensional sphere is the one-point compactification of $\mathbb{R}^3$ obtained by adding a point at infinity; we write $S^3 = \mathbb{R}^3 \cup \{\infty\}$. We will not define the one-point compactification, as that would take us too far afield. However, the reader can gain some intuition about $S^3$ by thinking of an ordinary two-dimensional sphere, $S^2$, as being obtained from the plane, $\mathbb{R}^2$, by adding a point at $\infty$. The interior of a disk is homeomorphic to $\mathbb{R}^2$ (the reader should verify this), so we can obtain $S^2$ by gluing all the points along the boundary of the disk together into a single point and calling that glued-up point $\infty$. Thus we imagine $D^2$ is like a drawstring bag that closes at the point $\infty$ to form $S^2$. Analogously, we can obtain $S^3$ by starting with a three-dimensional unit ball $B^3 = \{(x, y, z) \mid x^2 + y^2 + z^2 \leqslant 1\}$. The interior of this ball is homeomorphic to $\mathbb{R}^3$ (the reader should also verify this). Thus we obtain $S^3$ by gluing together all the points along the boundary of the three-dimensional ball into a single point and calling that glued-up point $\infty$. Of course the three-dimensional construction is somewhat harder to imagine than the two-dimensional construction, because we cannot do the gluing in $\mathbb{R}^3$. Although there are many ways to think of $S^3$, the most convenient way is to think of $S^3 = \mathbb{R}^3 \cup \{\infty\}$, because then we can work in $\mathbb{R}^3$ and add the point at $\infty$ only when it seems necessary. Also, starting with $S^3$ we can choose any point to label as $\infty$, and then by removing this point we obtain $\mathbb{R}^3 = S^3 - \{\infty\}$.

Adding a point at $\infty$ is somewhat disturbing for chemists, who tend to live in the real world, which they imagine to be $\mathbb{R}^3$ without any $\infty$. However, in most cases, when we prove a result about graphs in $S^3$, the same result will follow for graphs in $\mathbb{R}^3$. Whenever this is not the case we will say so explicitly. So, although we may do some proofs in $S^3$ rather than $\mathbb{R}^3$ because it is easier, in general our results will also be true in $\mathbb{R}^3$. For example, we have an analogous definition of ambient isotopy for graphs in $S^3$, and we can see as follows that a graph is ambient isotopic to its mirror image in $S^3$ if and only if it is ambient isotopic to its mirror image in $\mathbb{R}^3$. Let $A$ and $B$ be graphs in $\mathbb{R}^3$. On one hand, suppose there is an ambient isotopy $F : \mathbb{R}^3 \times I \to \mathbb{R}^3$ from a graph $A$ in $\mathbb{R}^3$ to a graph $B$ in $\mathbb{R}^3$. Now $\mathbb{R}^3$ is a subset of $S^3 = \mathbb{R}^3 \cup \{\infty\}$ so $A$ is also a graph in $S^3$, and, by defining $F(\infty, t) = \infty$ for all $t$, we can extend $F$ to an ambient isotopy from the graph $A$ to the graph $B$ as a subset of $S^3$. On the other hand, suppose that the graphs $A$ and $B$ are embedded in $S^3$ and there is an ambient isotopy $F$, in $S^3$, from the graph $A$ to the graph $B$. Observe that $F(A \times I)$ will not include every point of $S^3$. Thus there are points that $A$ does not pass through during the isotopy. We can modify $F$ a little, if necessary, to guarantee that some point that $A$ does not pass through during the isotopy is actually fixed throughout the isotopy. We label this point as $\infty$, and we then have $F(\infty, t) = \infty$ for all $t \in I$. Now we consider the graphs $A$ and $B$ as subsets of $\mathbb{R}^3 = S^3 - \{\infty\}$. Then we restrict $F$ to the set $(S^3 - \{\infty\}) \times I = \mathbb{R}^3 \times I$, and we will have an ambient isotopy from the graph $A$ to the graph $B$ that takes place in $\mathbb{R}^3$. Therefore, two graphs are ambient isotopic in $S^3$ if and and only if they are ambient isotopic in $\mathbb{R}^3$. Thus an embedded graph is topologically achiral in $S^3$ if and only if it is topologically achiral in $\mathbb{R}^3$. In particular, this means that we can prove theorems in $S^3$ and still have real results for chemists.

In addition to using the concept of ambient isotopy to define topological achirality, we can also express topological achirality in terms of homeomorphisms, which in some cases will be easier to work with. We begin by defining when two homeomorphisms are isotopic. The reader should compare this definition with the definition of when two subsets of $\mathbb{R}^3$ are ambient isotopic.

**Definition.** Let $A$ and $B$ be subsets of $\mathbb{R}^3$ or $S^3$, and let $h : A \to B$ and $g : A \to B$ be homeomorphisms. We say that $h$ and $g$ are *isotopic* if there exists a continuous function $F : A \times I \to B$ such that $F(x, 0) = h(x)$, $F(x, 1) = g(x)$, and for every fixed $t \in I$, the function $F(x, t)$ is a homeomorphism.

Intuitively, two homeomorphisms are isotopic if one can be continuously deformed into the other. An important result in topology is that every homeomorphism $h : \mathbb{R}^3 \to \mathbb{R}^3$ or $h : S^3 \to S^3$ is isotopic to either the identity map or to a reflection map, but not to both (for a proof see Alexander, 1930 and Cerf, 1968). We give these two types of homeomorphisms names as follows.

**Definition.** Let $h$ be a homeomorphism from $\mathbb{R}^3$ to itself or from $S^3$ to itself. If $h$ is isotopic to the identity map, then we say that $h$ is *orientation preserving*. If $h$ is isotopic to a reflection map, then we say that $h$ is *orientation reversing*.

Every homeomorphism of $\mathbb{R}^3$ or $S^3$ is either orientation preserving or reversing. An orientation-reversing homeomorphism maps an embedded graph to its mirror image, possibly deforming it as well, while an orientation-preserving homeomorphism maps an embedded graph to itself or to a deformation of the graph in space. A reflection map is orientation reversing, whereas any translation or rotation is orientation preserving. Observe that the composition of two reflections is orientation preserving. For example, when you see your reflection in two mirrors, your left and your right hands will not be reversed in the double reflection. It follows that the composition of two orientation-reversing homeomorphisms is an orientation-preserving homeomorphism. In contrast, the composition of an orientation-reversing homeomorphism, and an orientation-preserving homeomorphism is an orientation-reversing homeomorphism, and the composition of two orientation-preserving homeomorphisms is an orientation-preserving homeomorphism.

**Definition.** A graph $G$ that is embedded in $\mathbb{R}^3$ is *topologically achiral* if there exists an orientation-reversing homeomorphism of $(\mathbb{R}^3, G)$. If no such homeomorphism exists, then we say that $G$ is *topologically chiral*.

We can see that this definition is equivalent to the previous one as follows. If the graph $G$ is ambient isotopic to its mirror image by an isotopy $F$ of $\mathbb{R}^3$, then $F(x, 0)$ is the identity map of $\mathbb{R}^3$ and $F(G \times \{1\}) = G^*$, where $G^*$ denotes the mirror image of $G$. Hence, $F(x, 1)$ is a homeomorphism of $\mathbb{R}^3$ that is isotopic to the identity, and so $F(x, 1)$ is an orientation-preserving homeomorphism of $\mathbb{R}^3$ taking $G$ to its mirror image $G^*$. Now, by composing $F(x, 1)$ with a reflection map, we get an orientation-reversing homeomorphism of $\mathbb{R}^3$ that takes $G$ to itself. In contrast, suppose that we start with an orientation-reversing homeomorphism of $\mathbb{R}^3$ that takes $G$ to itself. The composition of this orientation-reversing homeomorphism together with a reflection map is an orientation-preserving homeomorphism of $\mathbb{R}^3$ that takes $G$ to its mirror image. Because any orientation-preserving homeomorphism of $\mathbb{R}^3$ is isotopic to the identity map, the homeomorphism of $\mathbb{R}^3$ taking $G$ to its mirror image $G^*$ must be isotopic to the identity. It follows that there must be an ambient isotopy from $G$ to $G^*$. Thus the two definitions of topological achirality are equivalent, and hence the two definitions of topological chirality are equivalent.

On one hand, to prove that an embedded graph is topologically achiral, we find it generally convenient just to exhibit a deformation of the graph to its mirror image by drawing a series of pictures. On the other hand, to prove that an embedded graph $G$ is topologically chiral, we will show that there cannot exist an orientation-reversing homeomorphism of $(\mathbb{R}^3, G)$. Note that the above definition could have been analogously stated in terms of homeomorphisms of $S^3$ rather than $\mathbb{R}^3$, and the definition of topological achirality in terms of homeomorphisms would still be equivalent to the one in terms of ambient isotopies.

## Knot and Link Types

Recognizing the symmetries of a completely rigid embedded graph is usually not that difficult. In contrast, it is often not at all obvious how to determine whether or not a flexible embedded graph is topologically chiral. Even using topology there is no algorithm to determine this, because different techniques will be useful in different situations. As we shall see, determining when a knot or a link is topologically chiral can often help determine when an embedded graph is topologically chiral. Because molecular knots and links can now be synthesized, it can also be chemically important to be able to determine when a knotted or linked structure is chiral. Thus, we will explore some techniques for determining when a knot or a link is topologically chiral, and then we return to a discussion of the chirality of embedded graphs.

**Definition.** A *simple closed curve* or a *knot* is defined as any embedding of $S^1$ in $\mathbb{R}^3$ or $S^3$. A *link* is the union of one or more disjoint knots; these knots are said to be the *components* of the link. Two knots or links are said to be *equivalent* if one is ambient isotopic to the other. A pair of knots or links that are equivalent are said to be of the same *knot type* or *link type*, respectively.

Note that this definition allows links to consist of only one component, so we will sometimes use the word *link* to refer to a knot or a link. A link is topologically achiral if and only if it and its mirror image are of the same link type. From a topology point of view, we are actually interested in knots and links only up to ambient isotopy. That is, we want to consider two knots to be different only if their knot types are different, whereas technically they are different if their embeddings are different. By an abuse of notation we shall henceforth use the terms *knot* and *link* to actually mean knot type or link type, respectively. For example, we shall call a knot a trefoil knot if it can be deformed to the usual trefoil knot, even if its precise embedding is not that of the usual trefoil.

**Definition.** The *unknot* or *trivial knot* is a simple closed curve in $\mathbb{R}^3$ (or $S^3$) that is ambient isotopic to a circle in a plane (or sphere). The *unlink* is a link in $\mathbb{R}^3$ (or $S^3$) that is ambient isotopic to a collection of disjoint circles in a plane (or sphere).

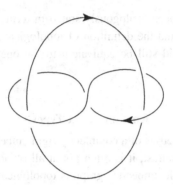

**Figure 2.2.** A projection of an oriented link

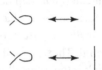

**Figure 2.3.** A kink can be added or removed

It is often convenient to work with diagrams of knots and links, that is, projections of the knots and links into a plane where each undercrossing is indicated by a gap in an arc. Figure 2.2 illustrates a projection of an oriented link. We project a link into a plane in such a way that at most two points of the link lie over a single point of the plane. We think of a projection as representing a link in $\mathbb{R}^3$, but in reality a projection is just a collection of disjoint arcs in the plane of the paper. The beginning and ending of each arc correspond to where a piece of the link goes under another piece. In Figure 2.2 the projection of the link is made up of five arcs.

A given knot or link has many different projections, and we may want to know if two projections represent the same knot or link. Two projections of a given knot or link may be different because the knot or link has been deformed or because we are projecting onto a different plane, or both. In 1926, Kurt Reidemeister (Reidemeister, 1926) proved that two knots or links are equivalent if and only if it is possible to get from a projection of one to a projection of the other by performing a finite sequence of moves from the following list.

0. Any arc of the projection can be deformed within the plane while keeping its endpoints fixed.
1. A kink can be added or removed as in Figure 2.3.

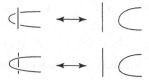

**Figure 2.4.** A strand can be slid over or under a strand

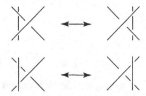

**Figure 2.5.** A strand can be slid over or under a crossing

2. A strand can be slid over or under another strand as in Figure 2.4.
3. A strand can be slid over or under a crossing as in Figure 2.5.

Although Reidemeister's result tells us that if two knots are equivalent we can get from a projection of one to a projection of the other by a sequence of these moves, it is often hard to find a particular sequence of moves that works. In addition, if we cannot find such a sequence of moves it does not necessarily mean that no such sequence exists. So, Reidemeister's Theorem is not practical for determining whether or not a given knot is topologically chiral.

## Link Polynomials

One important technique to detect chirality is to make use of a link polynomial. In general, these polynomials are associated with *oriented links*, that is, links with an arrow assigned to each component to indicate a particular direction. Such polynomials are powerful tools because they have the property that if one oriented link is ambient isotopic to another then the two will have the same polynomial. A numerical or algebraic expression associated with a link or an embedded graph is said to be a *topological invariant* if it is not changed by an ambient isotopy of the link or the graph. If a particular link polynomial is a topological invariant and if $L$ is an oriented link that is ambient isotopic to its mirror image, then the polynomial of $L$ and the polynomial of its mirror image $L^*$ will be the same. If these polynomials are not the same, then right away we will know that $L$ is topologically chiral.

*2. Detecting Chirality*

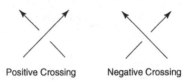

Positive Crossing        Negative Crossing

**Figure 2.6.** A positive crossing and a negative crossing

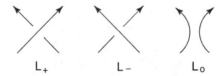

**L₊**            **L₋**            **L₀**

**Figure 2.7.** $L_+$, $L_-$, and $L_0$ are identical except at this crossing

There are several useful link polynomials, including the Jones polynomial (Jones, 1985), the Kauffman polynomial (Kauffman, 1987), and the two-variable HOMFLY polynomial (Freyd et al., 1985). These polynomials are somewhat similar in flavor, so we present only one. The polynomial that we will present is the HOMFLY polynomial, named after five of the authors who discovered it, Hoste, Ocneanu, Millet, Freyd, Lickorish, and Yetter (Freyd et al., 1985), though it was also independently discovered by Przytycki and Traczyk (1987). All of the link polynomials are actually Laurent polynomials, which means that the variables in the polynomials can be raised to negative as well as positive powers.

In order to define the HOMFLY polynomial (also known as the *P-polynomial*) we first orient $L$; then we fix a particular projection of $L$. The P-polynomial of $L$ will be defined in terms of the crossings of this oriented projection. We want to distinguish two different types of oriented crossings, which we will call positive crossings and negative crossings. A positive crossing corresponds to a right-handed twist and a negative crossing corresponds to a left-handed twist. These two types of crossings are illustrated in Figure 2.6.

The P-polynomial $P(L)$ has variables $m$ and $l$ and is defined from the oriented projection of $L$ by using the following three axioms.

1. $P(\text{unknot}) = 1$
2. Suppose $L_+$, $L_-$, and $L_0$ are oriented link projections that are identical except near a single crossing, where they differ by a positive, negative, or null crossing, respectively (illustrated in Figure 2.7). Then $l\,P(L_+) + l^{-1}P(L_-) + m\,P(L_0) = 0$.
3. $P(L)$ is not changed by an ambient isotopy of $L$.

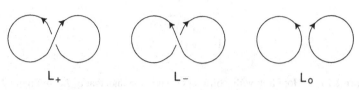

**Figure 2.8.** $L_+$, $L_-$, and $L_0$

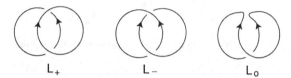

**Figure 2.9.** An oriented Hopf link is represented by $L_-$

This polynomial appears to depend on the particular projection of the link that is chosen. However, the P-polynomial was proven to be both well defined and uniquely determined by the three axioms (Freyd et al., 1985).

Given an oriented link, we compute its P-polynomial in terms of the polynomials of simpler links, which in turn are computed in terms of the polynomials of links that are simpler still, and so on until we get an unknot whose polynomial is known to equal 1.

We illustrate how to compute the P-polynomial by considering a couple of examples. First let $L$ consist of the oriented unlink of two components, which is represented by $L_0$ in Figure 2.8. Then $L_+$ and $L_-$ are as shown in the figure.

Now we use the axioms of the definition of the P-polynomial, together with the observation that both $L_+$ and $L_-$ are ambient isotopic to the unknot, in order to obtain the equation $l + l^{-1} + mP(L) = 0$. Hence $P(L) = -m^{-1}(l + l^{-1})$. Note that $L$ was the disjoint union of two unknots, yet the P-polynomial of $L$ was not equal to the number 2.

We shall use the P-polynomial of a two-component unlink to enable us to compute the P-polynomial of the oriented Hopf link, illustrated as $L_-$ in Figure 2.9. We choose the upper crossing to change so that we have $L_+$ and $L_0$ as indicated in Figure 2.9.

Now we see that $L_0$ is ambient isotopic to the unknot and $L_+$ is ambient isotopic to the link whose polynomial we computed above. By Axiom 3, we can substitute this polynomial into the equation given by Axiom 2 to get the equation $l(-m^{-1}(l + l^{-1})) + l^{-1}P(L_-) + m = 0$. Thus, after some simplification we get $P(L_-) = l^3 m^{-1} + l m^{-1} - l m$.

**Figure 2.10.** A Hopf link with a different orientation than that of $L_-$ in Figure 2.9

**Figure 2.11.** A positive crossing in a knot is independent of the orientation of the knot

It can be seen from these simple examples that computing the P-polynomial of any complicated knot or link will be quite cumbersome. However, there are a number of excellent computer programs that will compute all of the link polynomials for any link drawn with up to approximately fifty crossings (see, e.g., the programs *KnotTheoryByComputer*, Ochiai & Yamada, 1992 and *Knotscape*, Hoste & Thistlewaite, 1998). Peter Suber has prepared a list of current resources on knot theory including software programs for computing knot polynomials (see http://www.earlham.edu/~peters/knotlink.htm).

Notice that the orientation of the components of a link may affect its P-polynomial. For example, the P-polynomial of the Hopf link that is oriented as in Figure 2.10 turns out to equal $l^{-3}m^{-1} + l^{-1}m^{-1} - l^{-1}m$, which is different than the P-polynomial that we computed above for the oriented Hopf link in Figure 2.9. Specifically, the roles of $l$ and $l^{-1}$ have been reversed in the P-polynomial of the link in Figure 2.10 relative to the P-polynomial of the Hopf link illustrated as $L_-$ in Figure 2.9.

In contrast with links, we can see as follows that the orientation of a knot has no effect on its P-polynomial. Suppose that $L$ is a knot. Then reversing the orientation of $L$ has the effect of reversing all of the arrows in the projection. In particular, it reverses the direction of both arrows occurring at any crossing. As can be seen from the positive crossing illustrated in Figure 2.11, a positive crossing remains positive after both arrows are reversed; it is simply rotated by 180°. Negative crossings and null crossings also remain unchanged. Thus the P-polynomial of a knot $L$ is independent of the orientation of the knot. If we reversed the orientation of every component of a link, the P-polynomial also would not change. However, each component of a link can be oriented in two different ways. Thus we can change the orientation of one component without changing the orientation of every component, and this may affect the P-polynomial, as it did for the Hopf link.

Link polynomials can be quite useful for distinguishing knots and oriented links, because two knots or oriented links with different polynomials cannot be ambient isotopic. However, the link polynomials do not distinguish every pair of knots or links. In fact, Kanenobu (1986) has shown that there are infinitely many knots with the same P-polynomial.

The following theorem will tell us how to conclude from the polynomial of an oriented link that the oriented link is topologically chiral. We have defined topological chirality of links but not of oriented links, so, for clarity, we define an oriented link $L$ to be *topologically achiral* if there exists an orientation-reversing homeomorphism $h : (S^3, L) \to (S^3, L)$ that preserves the orientation of every component of $L$. If there is no such homeomorphism, then the oriented link is said to be *topologically chiral*. Equivalently, an oriented link $L$ is topologically achiral if there is an ambient isotopy taking the oriented link $L$ to the correspondingly oriented mirror image link $L^*$.

**Theorem 2.1.** *Let* L *be an oriented link with* P-*polynomial* P(L). *Let* $\bar{P}$(L) *denote the polynomial obtained from* P(L) *by interchanging* $l$ *and* $l^{-1}$. *If* P(L) $\neq$ $\bar{P}$(L) *then* L *is topologically chiral as an oriented link. If* L *is a knot and* P(L) $\neq$ $\bar{P}$(L) *then* L *is topologically chiral, independent of the orientation of* L.

*Proof.* Let $L^*$ denote the mirror image of the oriented link $L$. Then $L^*$ is obtained by reversing all of the crossings of $L$. That is, $L^*$ has positive crossings where $L$ has negative crossings and $L^*$ has negative crossings where $L$ has positive crossings. So, by progressively changing the crossings of $L^*$ as we compute its polynomial, the roles of $L^*_+$ and $L^*_-$ are the reverse of those of $L_+$ and $L_-$ in the computation of the polynomial for $L$. It follows from Axiom 2 of the definition of the P–polynomial that, for $L^*$ and all of the simpler links obtained from $L^*$, the roles of $l$ and $l^{-1}$ are the opposite of what they were for $L$ in the equation $lP(L_+) + l^{-1}P(L_-) + mP(L_0) = 0$. Hence in $P(L^*)$, the roles of $l$ and $l^{-1}$ have been interchanged relative to what they were in $P(L)$. Thus $P(L^*) = \bar{P}(L)$. We know that the P-polynomial is a topological invariant, so if there is an ambient isotopy from the oriented link $L$ to the oriented link $L^*$, then $P(L) = P(L^*) = \bar{P}(L)$. Thus if $P(L) \neq \bar{P}(L)$ then $L$ must be topologically chiral as an oriented link. Furthermore, if $L$ is a knot, then because the orientation has no effect on the polynomial of a knot, $P(L) \neq \bar{P}(L)$ implies that the knot $L$ is topologically chiral, independent of its orientation. □

As an example, the P-polynomial of the oriented Hopf link $L$ from Figure 2.9 was $P(L) = l^3m^{-1} + lm^{-1} - lm$. Because $P(L) \neq \bar{P}(L)$, Theorem 2.1 tells us

**Figure 2.12.** These oriented Hopf links are ambient isotopic

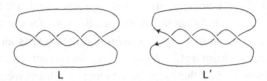

**Figure 2.13.** The unoriented and oriented (4, 2)-torus link

that the oriented Hopf link is topologically chiral. The mirror image $L^*$ of the oriented Hopf link $L$ of Figure 2.9 (reflected through the plane of the paper) is drawn on the left side of Figure 2.12. If we turn over the left-hand component of this link we obtain the oriented Hopf link of Figure 2.10, which is illustrated on the right side of Figure 2.12. Thus $L^*$, the mirror image of the oriented Hopf link of Figure 2.9, is ambient isotopic to the oriented Hopf link of Figure 2.10. So the proof of Theorem 2.1 tells us that the oriented Hopf link of Figure 2.10 should have P-polynomial $P(L^*) = \bar{P}(L)$, which is precisely what we observed when we compared the P-polynomials of the two oriented Hopf links.

In contrast, as an unoriented link the Hopf link is topologically achiral. If we ignore the orientation on the links in Figure 2.12 we see that the unoriented Hopf link is ambient isotopic to its mirror image.

We can also use the P-polynomial to prove that certain unoriented links are topologically chiral. For example, let $L$ denote the link that is illustrated on the left in Figure 2.13. This type of link is called a *torus link* because it can be drawn on the surface of a torus without any self-intersections. This particular link is called a (4, 2)-torus link, because, when it lies on the torus, it twists four times around the torus in one direction (corresponding to the four crossings), while wrapping two times around the torus the other way (corresponding to the two strands). We cannot compute the P-polynomial of an unoriented link, so we let $L'$ denote the oriented link that is illustrated on the right side of Figure 2.13. The P-polynomial of $L'$ is $P(L') = -l^{-5}m^{-1} - l^{-3}m^{-1} + ml^{-5} - m^3l^{-3} + 3ml^{-3}$.

We create an unoriented link $L^*$, which is the mirror image of the unoriented link $L$, by switching all of the crossings of $L$. We shall prove that the unoriented link $L$ is not ambient isotopic to its mirror image $L^*$ by contradiction. Suppose that the unoriented link $L$ could be deformed to its mirror image. Hence the

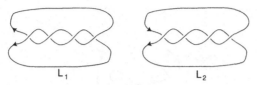

**Figure 2.14.** We orient the link $L^*$ in two different ways to get $L_1$ and $L_2$

**Figure 2.15.** A trefoil knot is topologically chiral

oriented $(4, 2)$-torus link $L'$ could be deformed to $L^*$ with some orientation. Because $L^*$ has two components, there are four ways that we could orient $L^*$. Two of these orientations for $L^*$ are illustrated as $L_1$ and $L_2$ in Figure 2.14. We can see that $L_1$ is the mirror image of $L'$. There are two additional ways of orienting $L^*$; however, each one of these will reverse the direction of both arrows at each crossing relative to one of the oriented links in Figure 2.14 and so will have the same P–polynomial as either $L_1$ or $L_2$. We compute the P–polynomial of the oriented links in Figure 2.14 to be $P(L_1) = -l^5 m^{-1} - l^3 m^{-1} + ml^5 - m^3 l^3 + 3ml^3$ and $P(L_2) = -m^{-1} l^{-3} - ml^{-1} - m^{-1} l^{-5} + ml^{-3}$. Because we are assuming that $L'$ is ambient isotopic to $L^*$ with some orientation, the P–polynomial of $L'$ should equal the P-polynomial of either $L_1$ or $L_2$. As this is not the case, we can conclude that the unoriented $(4, 2)$-torus link is topologically chiral.

For knots, we can apply Theorem 2.1 without the complication of dealing with orientations. For example, the trefoil knot, which is drawn in Figure 2.15, has P-polynomial $-l^4 + l^2 m^2 - 2l^2$, independent of how the knot is oriented. This polynomial is clearly not symmetric with respect to $l$ and $l^{-1}$, so we can immediately conclude that the trefoil knot is topologically chiral. With the help of one of the computer programs to compute link polynomials, we can use Theorem 2.1 to recognize many topologically chiral knots and oriented links.

Theorem 2.1, however, does not detect all topologically chiral knots and oriented links, because there are topologically chiral knots and oriented links whose P-polynomials are symmetric with respect to $l$ and $l^{-1}$. For example, the knot $9_{42}$, which is illustrated in Figure 2.16, has this property. Recall that the notation $9_{42}$ means that among the knots that can be drawn with nine crossings but no fewer, this is the forty-second such knot listed in the

**Figure 2.16.** Knot $9_{42}$ is topologically chiral

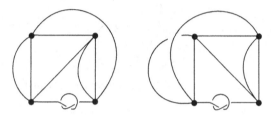

**Figure 2.17.** Rigid vertex embedded graphs that are not ambient isotopic

knot tables (see Rolfsen, 1976). The P-polynomial of this knot is $P(9_{42}) = (-2l^{-2} - 3 - 2l^2) + m^2(l^{-2} + 4 + l^2) - m^4$, which is symmetric with respect to $l$ and $l^{-1}$. However, with the use of the concept of the signature (for the definition see Murasugi, 1993) it is possible to show that this knot is topologically chiral.

## Chirality of Embedded Graphs

Growing out of the knot and link polynomials, several polynomial invariants have been developed for embedded graphs (Kauffman, 1989; Yamada, 1989; Yokota, 1996). Some of these are restricted to graphs, all of whose vertices have a valence of three or all of whose vertices have a valence of four. In most cases, the polynomials for embedded graphs with a valence of four are actually only useful for the so-called rigid vertex graphs. A *rigid vertex* graph is one in which each vertex is replaced by a two-dimensional disk with the edges attached to the boundary of the disk. The order of the edges around the boundary of the disk is fixed by any ambient isotopy of the rigid vertex graph because each edge must remain attached to the boundary of the disk throughout the ambient isotopy. For example, consider the rigid vertex embedded graphs of Figure 2.17. We imagine the black dots as small black disks. If the black dots represented vertices, then the two ordinary graphs would be ambient isotopic, by interchanging the positions of the inner and outer diagonals of the second picture. However, this ambient

**Figure 2.18.** To obtain $C(G)$ we replace each vertex disk by the four pictures on the right

**Figure 2.19.** We shall apply Theorem 2.2 to this rigid vertex embedded graph

isotopy changes the order of the edges around the vertices, and there is no corresponding ambient isotopy of the rigid vertex graphs.

A rigid vertex graph $G$ embedded in $\mathbb{R}^3$ is said to be *topologically achiral* if $G$ is ambient isotopic to its mirror image; otherwise $G$ is said to be *topologically chiral*. In particular, if $G$ is a topologically achiral rigid vertex graph then the ambient isotopy from $G$ to its mirror image must maintain the order of the edges emanating from each vertex disk.

Although Kauffman and others have generalized link polynomials to define polynomial invariants of these rigid vertex four-valent embedded graphs, Kauffman (1989) also defines a simpler invariant that he asserts is just as effective at detecting topological chirality for rigid vertex embedded graphs as any of the polynomials. We present Kauffman's invariant as follows.

For an embedded four-valent rigid vertex graph $G$, we define $C(G)$ to be the collection of knots and links associated with $G$ by successively replacing all of the vertex disks, together with the four strings emanating from them, by each one of the four pictures on the right hand side of Figure 2.18.

Kauffman (1989) proved that the set $C(G)$ is a topological invariant of any embedded rigid vertex graph $G$. From this result the theorem below follows.

**Theorem 2.2.** *Let G be a four-valent rigid vertex graph embedded in $\mathbb{R}^3$. If there is an element of C(G) that is topologically chiral and that cannot be deformed to the mirror image of any other element of C(G), then G is topologically chiral as a rigid vertex graph.*

To see how Theorem 2.2 can be used, we apply it to the embedded rigid vertex graph illustrated in Figure 2.19.

**Figure 2.20.** $C(G)$ for the rigid vertex graph in Figure 2.19

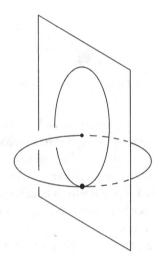

**Figure 2.21.** This graph is its own mirror image

The elements of the set $C(G)$ are illustrated in Figure 2.20. Because $C(G)$ contains a trefoil knot but not its mirror image, we can conclude that the rigid vertex graph $G$ is topologically chiral.

We can use the above example to make the distinction between the topological chirality of rigid vertex graphs and the topological chirality of ordinary embedded graphs. If we replace the vertex disk in Figure 2.19 by an ordinary vertex, this graph turns out to be topologically achiral. The ordinary graph is illustrated in a symmetry presentation in Figure 2.21, where there is a vertical mirror plane that reflects the graph to itself. Because this graph is its own mirror image (without requiring any flexibility), if it were a molecular bond graph with the vertex representing an atom and the two circles representing molecular chains, then the molecule would be chemically achiral. Although the order of bonds around an atom may be rigidly fixed in three-dimensional space, they are generally not rigidly fixed around a two-dimensional disk.

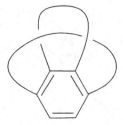

**Figure 2.22.** The benzene ring in this graph acts as a rigid vertex disk

**Figure 2.23.** We shall compute $T(G)$ for this embedded graph $G$

In contrast, if we create a molecule in the form of Figure 2.19 but in which the vertex disk is replaced by a hexagon representing a benzene ring, then the bonds attaching the benzene ring to the rest of the molecule occur in a fixed order around the hexagon (see Figure 2.22). Thus the benzene ring would act as a rigid vertex disk, and as such the graph would be topologically chiral. Furthermore, the molecule represented by such a graph would be chemically chiral. Hence, for certain molecules it makes sense to use rigid vertex graphs and for others it does not.

We return now to our study of the topological chirality of ordinary embedded graphs. Yokota (1996) has introduced a knot polynomial invariant for arbitrary embedded graphs; however, this polynomial cannot be used to detect chirality because it cannot distinguish between an embedded graph and its mirror image. In contrast, Kauffman (1989) has defined a topological invariant for ordinary embedded graphs that is similar to his invariant for rigid vertex graphs and that can be used to detect chirality. We shall describe Kauffman's invariant below. Let $G$ be an embedded graph, and define $T(G)$ to be the set of all knots and links contained in the graph $G$. That is, the set $T(G)$ contains every knot and link that can be obtained by joining any two edges together at each vertex and disconnecting all other edges from that vertex. This operation creates a number of arcs that are not closed up, but $T(G)$ does not include these arcs. For example, consider the embedded graph $G$ illustrated in Figure 2.23.

*2. Detecting Chirality*

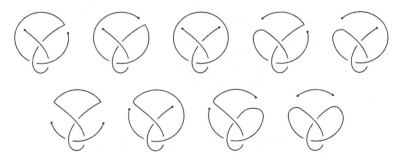

**Figure 2.24.** The arcs and simple closed curves that we get from *G*

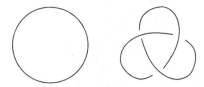

**Figure 2.25.** The elements of $T(G)$ for the graph in Figure 2.23

The total collection of arcs and simple closed curves that we get by performing this operation on the graph in Figure 2.23 is illustrated in Figure 2.24; the elements of the set $T(G)$ are illustrated in Figure 2.25.

Kauffman (1989) proved that, for any embedded graph $G$, the set $T(G)$ is invariant up to ambient isotopy. The theorem below follows.

**Theorem 2.3.** *Let G be a graph embedded in* $\mathbb{R}^3$. *If there is an element of* T(G) *that is topologically chiral and that cannot be deformed to the mirror image of any other element of* T(G), *then G is topologically chiral.*

We can apply Kauffman's Theorem to the graph $G$ that is illustrated in Figure 2.23. As can be seen from Figure 2.25, the only nontrivial knot that $T(G)$ contains is the right-handed trefoil knot. This knot is topologically chiral, so $G$ is topologically chiral by Theorem 2.3.

### Chirality of Molecular Knots and Links

We shall now apply the techniques we have presented to prove the topological chirality of some molecular knots and links. Recall that if we succeed in proving that a molecular graph is topologically chiral, then it will follow that

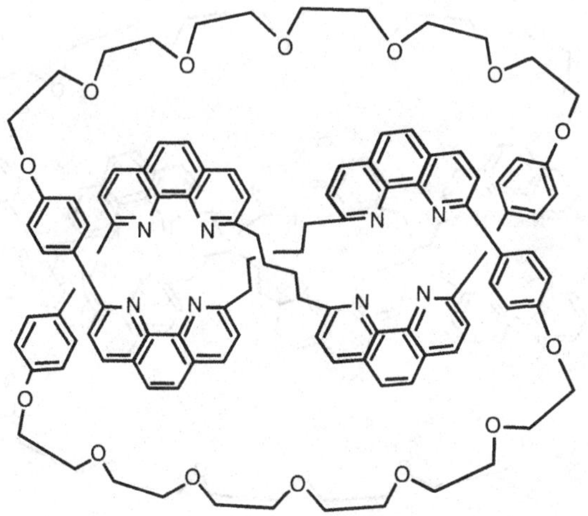

**Figure 2.26.** The molecular trefoil knot

the molecule that it represents is chemically chiral, because any molecular motion corresponds to a rigid, flexible, or partially flexible deformation of the molecular graph.

First we consider the molecular trefoil knot synthesized by Dietrich-Buchecker and Sauvage (1989). We illustrate this knot in Figure 2.26. If we imagine that each of the hexagons in the molecular graph has been replaced by a single edge, then we would obtain an ordinary right-handed trefoil knot. As we proved already that the right-handed trefoil knot is topologically chiral, it is tempting to conclude from this observation that the embedded graph of the molecular trefoil knot must also be topologically chiral. However, the graph of the molecular trefoil knot has hexagonal rings, which are topologically quite different from the single edges in the trefoil knot. If we want to make sure that our conclusions are mathematically correct, then we cannot just replace these hexagonal rings by single edges.

In order to formally prove the topological chirality of the molecular trefoil knot, we can use Kauffman's Theorem. We create the set $T(G)$ of all the knots and links contained in the graph $G$ of the molecular trefoil knot. Every non-trivial knot contained in $G$ goes around part of each hexagonal ring. There are two ways to go around a single hexagonal ring, so we can see that there are many knotted simple closed curves contained in $G$. However, every one of these knots is a right-handed trefoil knot. Thus, $T(G)$ consists of a large collection of unlinks and right-handed trefoil knots, and nothing else. Because a right-handed trefoil knot is topologically chiral and $T(G)$ does not contain any left-handed

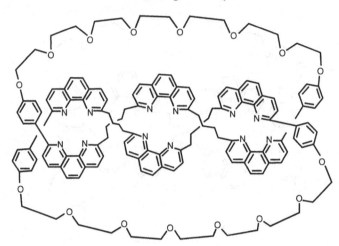

**Figure 2.27.** The molecular (4, 2)-torus link

trefoil knots, we can conclude from Kauffman's Theorem that the embedded graph $G$ is topologically chiral.

We can use this same approach to prove that other molecular knots and links are topologically chiral. For example, consider the molecular link illustrated in Figure 2.27. This catenane was synthesized by Nierengarten, Dietrich-Buchecker, and Sauvage (1994). For this molecule, the set $T(G)$ consists of many unlinks together with many copies of the (4, 2)-torus link, which was illustrated as $L$ in Figure 2.13. The set $T(G)$ does not contain the mirror image of the (4, 2)-torus link. We saw earlier that this unoriented link is topologically chiral. Therefore, the molecular (4, 2)-torus link is topologically chiral as well.

If we want to prove that an oriented molecular link is topologically chiral, then applying Kauffman's Theorem becomes somewhat more complicated. For example, Mitchell and Sauvage (1988) synthesized the oriented Hopf link, which is illustrated on the left in Figure 2.28. As we shall see shortly, each group of three hexagons with another hexagon as a tail orients the component of the link that contains it. However, unless we modify it, Kauffman's method will neglect this information; and $T(G)$ will contain only unlinks and unoriented Hopf links. This will not help us, because an unoriented Hopf link is topologically achiral. In order to avoid this problem, we shall consider the molecular graph in Figure 2.28 to be a labeled graph, where vertices are labeled N or O as indicated (as usual, we do not label the carbons), and in addition we put an X at the vertices where the tail is attached in order to keep track of this information. The graph on the right in Figure 2.28 is obtained by adding these two X labels.

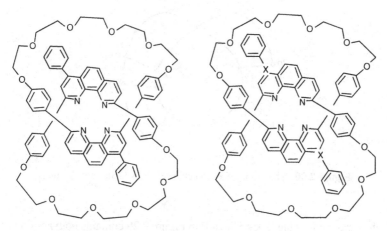

**Figure 2.28.** The oriented molecular Hopf link

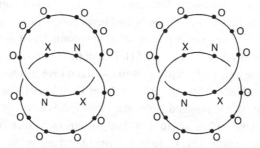

**Figure 2.29.** The labeling of this Hopf link gives it an orientation, which makes it topologically chiral

Adding these labels helps us keep track of the orientations while we are applying Kauffman's method. More specifically, if we could deform the embedded graph of the oriented molecular Hopf link to its mirror image, then the deformation would take the labeled graph to its mirror image, sending each N to an N, each O to an O, and each X to an X. If we let $G$ denote the embedded labeled graph that is illustrated on the right in Figure 2.28, then $T(G)$ contains many links and unlinks. All of the links that are not the unlink are Hopf links, and each component of these Hopf links contains six O's. However, not all components of these Hopf links contain an X, or the same number of N's, or the same number of unlabeled vertices (coming from carbon atoms). We do not list all of the possibilities, as there are quite a few. What is important is that some of the Hopf links in $T(G)$ look like the labeled link on the left in Figure 2.29; however, none of the links in $T(G)$ look like its mirror image, which is on the right of Figure 2.29.

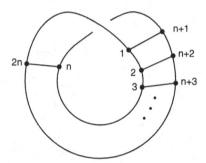

**Figure 2.30.** The graph of a molecular Möbius ladder with $n$ rungs

Each component of the link on the left in Figure 2.29 contains both an X and an N, in addition to the six Os. We can imagine that there is an arrow on the edge going from the X toward the adjacent N. If we could deform the link on the left in Figure 2.29 to its mirror image, illustrated on the right, this deformation would take each X to an X and each N to an N, and hence would preserve the direction of these arrows. Because we saw, using the P-polynomial, that an oriented Hopf link is topologically chiral, the labeled Hopf link that is illustrated on the left in Figure 2.29 is topologically chiral as well. As the mirror image of this labeled link is not contained in $T(G)$, it follows from Kauffman's Theorem that the embedded graph of the oriented molecular Hopf link is topologically chiral.

We can use these techniques to show that various molecular knots and links are topologically chiral. This includes not only synthetic molecular knots and links of the types shown, but proteins containing knots or links as well. See Liang and Mislow (1995a) for some examples of knotted and linked proteins.

For a complex graph, enumerating the set $T(G)$ may be quite tedious, and it may be necessary to use a link polynomial or another method to check the topological chirality of the knots and links contained in $T(G)$. However, by using a computer program to find $T(G)$ and then to compute the P-polynomials of the elements of $T(G)$, we find this invariant can be very useful for detecting topological chirality of embedded graphs. Because topological chirality implies chemical chirality, Kauffman's method is potentially useful for chemists.

In contrast, as with the P-polynomial, this embedded graph invariant will not detect the chirality of every topologically chiral embedded graph. For example, Figure 2.30 illustrates a projection of a molecular Möbius ladder with $n$ rungs. Since this projection has only one crossing, it follows that $T(M_n)$ will only contain the unknot and unlinks. So Theorem 2.3 will not help us establish topological chirality. However, in Chapter 3 we shall present some completely different techniques that demonstrate that this embedded graph is topologically chiral.

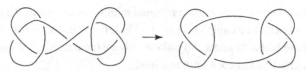

**Figure 2.31.** An alternating knot with a removable crossing

**Figure 2.32.** A projection is reduced if it does not contain these types of crossings

## Alternating Links

We have seen from the embedded graph invariants discussed above that being able to detect topologically chiral knots and links can be useful to prove the topological chirality of embedded graphs. We return now to our discussion of the topological chirality of knots and links and focus on a special class of links that are easier to deal with. The links in this class are called *alternating links*.

**Definition.** A projection of a link $L$ is said to be *alternating* if, as one traverses each component, the crossings alternate between over and undercrossings. Any link that has an alternating projection is said to be an *alternating link*.

For example, on one hand; the projection of knot $9_{42}$, illustrated in Figure 2.16, is not alternating. On the other hand, the two knot projections illustrated in Figure 2.31 are both alternating. Notice that the middle crossing in the projection on the left can easily be removed by turning the right side of the knot over. If we did this, then the projection would have fewer crossings, as illustrated on the right side of Figure 2.31.

In order to minimize the number of crossings, we wish to avoid any crossing that looks like the removable crossing on the left side of Figure 2.31.

**Definition.** A projection of a link is said to be *reduced* if it contains no crossings of the types illustrated in Figure 2.32. There can be additional crossings within the black disks, but no other strands of the link can enter the black disks.

Using link polynomials, Murasugi (1987) and Thistlethwaite (1987) independently proved the following theorem, which makes it easy to recognize when a projection of an alternating link has a minimal number of crossings. This theorem is quite powerful, because for nonalternating links it is often very difficult to know if a given projection has a minimal number of crossings. Knowing the minimal number of crossings of a knot or link helps us to classify knots and links, because tables of knots and links are organized according to the minimal number of crossings of each knot and link.

**Theorem 2.4.** *If a link has a reduced alternating projection, then that projection has the minimal number of crossings of any projection of that link.*

For example, the projection on the right side of Figure 2.31 is reduced alternating, and hence has a minimal number of crossings.

We shall now define the writhe of a projection, which can be used to detect topological chirality for alternating links.

**Definition.** Consider a projection of an oriented link $L$. We define the *writhe* of this projection as the sum of $+1$ for every positive crossing of the link (defined as in Figure 2.6 of this chapter) and $-1$ for every negative crossing of the link.

Notice that the writhe is not a topological invariant of links. For example, no matter how we choose the orientations of the knot projections in Figure 2.31, we see that one has writhe $-5$ and the other has writhe $-6$. But both of these projections represent the same knot. As with polynomials, the writhe of any knot projection is independent of the specific choice of orientation assigned to the knot. Thistlethwaite (1988) proved that the writhe of a reduced alternating projection is a topological invariant for alternating oriented links. In particular, he proved the following theorem.

**Theorem 2.5.** *All reduced alternating projections of an alternating oriented link have the same writhe.*

Theorem 2.5 implies that if an alternating oriented link is topologically achiral, then the writhe of its reduced alternating projection is zero. We see this as follows. Suppose that $L$ is an oriented link with a reduced alternating projection that has writhe $w$. The mirror image of this projection is a reduced alternating projection of the oriented link $L^*$, which is the mirror image of $L$. Notice that every positive crossing of the projection of $L^*$ corresponds to a negative crossing of the projection of $L$, and every negative crossing of the projection of $L^*$ corresponds to a positive crossing of the projection of $L$. It follows that the writhe of this projection of $L^*$ is $-w$. If $L$ is topologically achiral then $L$ is ambient isotopic to $L^*$. Thus $L$ and $L^*$ are reduced alternating projections of equivalent links. Now by Theorem 2.5, the writhe of the reduced alternating

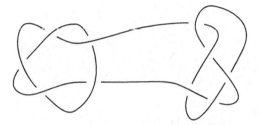

**Figure 2.33.** This knot is topological chiral but has zero writhe

projection of $L$ must equal the writhe of the reduced alternating projection of $L^*$. Hence $w = -w$, from which it follows that $w = 0$.

Because it is easy to recognize a reduced alternating projection by inspection, Theorem 2.5 gives us an easy way to recognize some oriented links that are topologically chiral. If we have a reduced alternating projection of an oriented link with nonzero writhe, we immediately conclude that the oriented link is topologically chiral. For example, we can apply this result to the right-hand picture of Figure 2.31. This is easily seen to be a reduced alternating projection that has a writhe equal to $-6$, hence this knot must be topologically chiral.

Of course, just because a reduced alternating projection of an oriented link has zero writhe does not necessarily mean that the oriented link is topologically achiral. For example, the knot illustrated in Figure 2.33 must be topologically chiral because its P-polynomial is not symmetric with respect to $l$ and $l^{-1}$, but it has a writhe equal to zero.

Also, just because the writhe of an oriented alternating link is nonzero does not imply that the unoriented link is topologically chiral. For example, we saw that the Hopf link is topologically achiral as an unoriented link, yet the reduced alternating projection illustrated in Figure 2.10 has a writhe of two. Next we will develop a way to use the writhe to help us detect topological chirality for unoriented alternating links, but we first need a definition.

**Definition.** Let $K_1$ and $K_2$ be oriented knot projections. We define the *linking number* of $K_1$ and $K_2$, written $\mathrm{Lk}(K_1, K_2)$, to be one-half of the sum of $+1$ for every positive crossing between $K_1$ and $K_2$ and $-1$ for every negative crossing between $K_1$ and $K_2$.

As an example consider the oriented links $9^2_{41}$ and $9^2_{61}$ illustrated on the left and the right, respectively, of Figure 2.34. The notation for links is similar to that for knots. In particular, $9^2_{41}$ refers to the fact that among all the two component links whose minimal crossing number is nine, this one is the forty-first one listed in the standard tables (see Rolfsen, 1976). Similarly, $9^2_{61}$ is the sixty-first

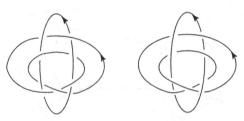

**Figure 2.34.** Oriented links $9^2_{41}$ and $9^2_{61}$

link of two components with minimal crossing number nine. The link $9^2_{41}$ is alternating, whereas $9^2_{61}$ is not alternating. The two components of the oriented link $9^2_{41}$ have linking number 0 because there are four positive crossings and four negative crossings between components, whereas the linking number of the components of the oriented link $9^2_{61}$ is $-4$ because all eight of the crossings between components are negative.

There are a number of important observations to make about the linking number. It is clear from our definition that $\text{Lk}(K_1, K_2) = \text{Lk}(K_2, K_1)$. In addition, because each component of the link must cross the other component an even number of times, the linking number is always an integer. Furthermore, it can be shown that the linking number is an invariant under the three Reidemeister moves illustrated at the beginning of this chapter. It follows from Reidemeister's Theorem that the linking number is invariant under an ambient isotopy of the oriented link $L = K_1 \cup K_2$. Thus; although the linking number was defined in terms of link projections, it actually is a topological invariant of any oriented two-component link.

This implies that if $L = K_1 \cup K_2$ is an oriented link, and $h$ is an orientation-preserving homeomorphism of $\mathbb{R}^3$, then $\text{Lk}(K_1, K_2) = \text{Lk}[h(K_1), h(K_2)]$. In contrast, any orientation-reversing homeomorphism $h : \mathbb{R}^3 \to \mathbb{R}^3$ will reverse all of the crossings of the link, so $\text{Lk}(K_1, K_2) = -\text{Lk}[h(K_1), h(K_2)]$. Finally, observe that reversing the direction of one of the components of the link interchanges all of the positive and negative crossings between this component and the other component. Hence, if we let a minus sign in front of a knot mean that the orientation of that knot has been reversed, then $\text{Lk}(K_1, K_2) = -\text{Lk}(-K_1, K_2) = -\text{Lk}(K_1, -K_2) = \text{Lk}(-K_1, -K_2)$.

In order to better understand how the writhe can help us detect topological chirality for reduced alternating projections of unoriented links, we will make use of the following definition.

**Definition.** Let $Q$ denote a projection of an oriented link. We define $i(Q)$ as the sum of $+1$ for every positive crossing between two strands of a single component of $Q$ and $-1$ for every negative crossing between two strands of a single component of $Q$, where the sum is taken over all of the components of $Q$.

Observe that if $Q$ is a knot projection then $i(Q)$ is just the writhe of the projection. If $Q$ is either of the links illustrated in Figure 2.34, then $i(Q) = 1$, because the only component that crosses itself has precisely one positive crossing. We have seen that the sign of a crossing between two strands of a single knot is unaffected by the orientation on the knot. Because $i$ is determined only by crossings of two strands of the same component, $i$ is independent of the orientation on any component of $Q$. We now prove the following theorem.

**Theorem 2.6.** *Let* L *be an unoriented alternating link, and let* $L_1$ *and* $L_2$ *each be an oriented reduced alternating projection of* L. *Then* $i(L_1) = i(L_2)$. *Furthermore, if the unoriented link* L *is topologically achiral, then* $i(L_1) = 0$ *and* $L_1$ *has an even writhe.*

*Proof.* We define $j(L_1)$ to equal the sum of two times the linking number of every pair of components of $L_1$. (Note that $L_1$ may have more than two components.) We can write $j(L_1) = \sum_{K_i, K_j \in L_1} 2 \operatorname{Lk}(K_i, K_j)$. We see from the definitions of the writhe and linking number that $i(L_1) = \text{writhe}(L_1) - j(L_1)$. That is, the algebraic sum of the crossing numbers of all crossings of $L_1$ minus the sum of the crossing numbers that occur between distinct components of $L_1$ is equal to the sum of the crossing numbers of crossings that occur within each individual component of $L_1$. Because the linking number of any two component oriented link is a topological invariant, by its definition, the value of $j$ is a topological invariant of any oriented link. Also by Theorem 2.5, any two reduced alternating projections of an oriented link have the same writhe, so it follows from the above equation that any two reduced alternating projections of an alternating oriented link will have the same value of $i$. Thus if $L_1$ and $L_2$ are both reduced alternating projections of $L$ with the same orientation, then we have $i(L_1) = i(L_2)$. However, as we observed above, the orientation of a link does not affect the value of $i$. It follows that, independent of their orientations, $i(L_1) = i(L_2)$.

Now assume there exists an ambient isotopy of the unoriented link $L$ to its mirror image. Let $L_1$ denote a fixed reduced alternating oriented projection of $L$ and let $L_3$ denote the mirror image of the oriented projection $L_1$. Because we are assuming that $L$ is ambient isotopic to its mirror image, $L_3$ is an oriented reduced alternating projection of $L$ after an ambient isotopy. Hence it follows by the above paragraph that $i(L_1) = i(L_3)$. In contrast, since $L_3$ is obtained by interchanging all of the positive and negative crossings of $L_1$, it follows that

$i(L_3) = -i(L_1)$. From these equations we deduce that $i(L_3) = -i(L_3)$, and so $i(L_1) = i(L_3) = 0$. This means that the writhe of $L_1$ is equal to $j(L_1)$. By its definition $j$ is always an even number, so the writhe of $L_1$ is even.                    □

Any link that is topologically achiral as an oriented link is also topologically achiral as an unoriented link. Recall that if $L$ is an alternating link that is topologically achiral as an oriented link, then it follows from Theorem 2.5 that the writhe of any reduced alternating projection of $L$ equals zero. Therefore, it follows from Theorem 2.6 and its proof that if $L_1$ is a reduced alternating projection of an oriented alternating link $L$ that is topologically achiral, then $i(L_1) = 0$ and hence $j(L) = 0$. Thus the sum of all the linking numbers between all of the pairs of distinct components of $L$ is zero. So if this sum is not zero, then we know that the link is topologically chiral as an oriented link. This does not work for unoriented links, because the Hopf link is an example of an unoriented alternating link that is topologically achiral but has a nonzero linking number between its only two components regardless of their orientations. For both oriented and unoriented alternating links, the following corollary provides a way to detect chirality.

**Corollary 2.7.** *Any topologically achiral alternating unoriented link has an even minimal crossing number.*

*Proof.* By Theorem 2.4, any reduced alternating projection of a link has a minimal crossing number. By Theorem 2.6, if the link is topologically achiral then the writhe of any reduced alternating projection must be even. So any reduced alternating projection must either have an odd number of positive crossings and an odd number of negative crossings, or an even number of positive crossings and an even number of negative crossings. This implies that the total number of crossings of this projection must be even.                    □

It is not hard to recognize a reduced alternating projection, so using Corollary 2.7 in conjunction with Theorem 2.4 is an easy way to prove that certain knots and links are topologically chiral without doing any computation. For example, Figure 2.35 shows a reduced alternating projection of the link $9^2_{41}$. We immediately conclude that this unoriented link is topologically chiral because this projection is reduced alternating and has an odd crossing number.

We can also sometimes use Theorem 2.6 to prove that an alternating unoriented link is chiral even when it has an even crossing number. For example, consider the three component, eight crossing link, $8^3_3$, which is drawn in Figure 2.36. The projection given in the figure can be seen to be reduced alternating.

**Figure 2.35.** A reduced alternating projection of the link $9_{41}^2$

**Figure 2.36.** Link $8_3^3$ is topologically chiral but has an even minimal crossing number

Observe that, no matter how the components of this link are oriented, the value of $i$ will be 2. Hence by Theorem 2.6, the unoriented link $8_3^3$ is topologically chiral.

## Tait's Conjecture

In 1898, in his papers on knot theory, Tait conjectured that every topologically achiral knot or link must have even minimal crossing number (Tait, 1898). Corollary 2.7 shows that Tait was right for alternating knots and links. However, it was shown by Liang and Mislow (1995b) and by Liang et al. (1998) that there are nonalternating links that are topologically achiral, yet have an odd minimal crossing number. We present such examples below. Consider the two-component link, $9_{61}^2$, which was illustrated in Figure 2.34. This is not an alternating link. Because this link has been classified in the tables, it is known that nine is the minimal number of crossings of any projection of this link up to ambient isotopy.

However, the link $9_{61}^2$ is ambient isotopic to the link illustrated in Figure 2.37, if we disregard the orientations on the components. The link in this projection can be thought of as the boundary components of two linked Möbius strips,

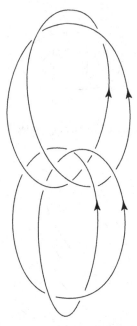

**Figure 2.37.** A symmetry presentation for the unoriented link $9^2_{61}$

where one Möbius strip has a right-handed half-twist and the other has a left-handed half-twist. After some examination, the reader will see that if we ignore the orientations, then this projection can be rotated by 90° about a vertical axis to obtain its mirror image through a horizontal plane cutting through the center of the picture. Thus $9^2_{61}$ is topologically achiral as an unoriented link and hence is a counterexample to Tait's conjecture for nonalternating links.

Observe from Figure 2.37 that this rotation takes the link to its mirror image, reversing the orientation of one of the components. In contrast with its topological achirality as an unoriented link, Doll and Hoste (1991) have shown that $9^2_{61}$ is topologically chiral as an oriented link. In fact, Doll and Hoste proved that no link with an odd minimal crossing number less than ten is topologically achiral as an oriented link.

Now consider the eleven crossing projection of a link, which is illustrated in Figure 2.38. This link, which we shall call $L$, is ambient isotopic to the link illustrated in Figure 2.39 if we ignore the orientations. If we rotate the link projection in Figure 2.39 by 90° about an axis perpendicular to the page going through the center of the picture, we will obtain the mirror image of the link. Furthermore, the reader can see from Figure 2.39 that the orientation of the link is preserved by this rotation. Thus $L$ is topologically achiral as an oriented link.

**Figure 2.38.** An eleven crossing link projection

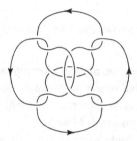

**Figure 2.39.** The link in Figure 2.38 is ambient isotopic to this projection, which can be rotated to its mirror image

It was proved by Liang et al. (1998) that the minimal crossing number of this link is eleven. This can also be seen from the recent knot and link tables of Hoste and Thistlethwaite (1998). Furthermore, Hoste, Thistlethwaite, and Weeks (1998) have found a nonalternating knot with a minimal crossing number of fifteen that is topologically achiral. This knot is illustrated in Figure 2.40. Thus Tait's conjecture that every topologically achiral link has an even minimal crossing number is false for both nonalternating knots and nonalternating links.

After considering the role of the crossing number in detecting topological chirality, it is natural to wonder whether the number of components of a link can also help detect topological chirality. In fact, Corinne Cerf (1997) proved that every oriented alternating link with an even number of components is topologically chiral (this result can also be obtained by using different methods from the work of Conway, (1967), and Crowell, (1959)). This result provides us with a huge class of topologically chiral oriented links. For example, any two-component alternating oriented link is topologically chiral. Of course, there are alternating oriented links with an odd number of components that are nonetheless topologically chiral. For instance, we saw above that unoriented three-component link $8_3^3$ was topologically chiral; hence as an oriented link it is also topologically

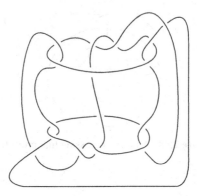

**Figure 2.40.** A fifteen crossing knot that is topologically achiral

chiral. Furthermore, counting the number of components will not enable us to prove that an unoriented link or a nonalternating link is topologically chiral. For example, the Hopf link is an alternating link with two components, yet it is topologically achiral as an unoriented link.

At present, there is no single method that will determine for every knot, link, or embedded graph whether or not it is topologically chiral. However, by combining various methods according to whether or not a link is oriented or alternating, we can often determine whether or not it is chiral. These results can in turn help us determine whether or not certain embedded graphs are topologically chiral, and if a molecular graph is topologically chiral then the molecule it represents is necessarily chemically chiral.

## Exercises

1. What does it mean to say that a link is topologically chiral as an oriented link versus as an unoriented link?
2. How can we use the P-polynomial to prove that an unoriented link is topologically chiral?
3. How can we use the P-polynomial to prove that an unoriented knot is topologically chiral?
4. How can we use the Reidemeister moves to prove that something is a topological invariant?
5. Is the writhe a topological invariant? Explain.
6. Is the minimal crossing number a topological invariant? Explain.
7. When can the writhe be used to show that a link is topologically chiral?
8. When can the minimal crossing number be used to show that a link is topologically chiral?
9. What is $i(L)$, and when can it be used to show that a link is topologically chiral?
10. Prove that if every vertex of a graph $G$ has a valence of at least three, then every homeomorphism $h : G \to G$ takes vertices to vertices and edges to edges. Give an

example of a graph $G$ and a homeomorphism $h : G \to G$, where $h$ does not take vertices to vertices and edges to edges.

11. Prove that the graph of the parabola $y = x^2$ is homeomorphic to the $x$-axis, both as subsets of $\mathbb{R}^2$.

12. Let $A = \{(x, 0, 0) \mid x \in \mathbb{R}\}$ and let $B = \{(x, x^2, 0) \mid x \in \mathbb{R}\}$. Prove that the sets $A$ and $B$ are ambient isotopic as subsets of $\mathbb{R}^3$.

13. Describe what the sets $S^1 \times S^1 \times I$, $S^1 \times S^1 \times S^1$, and $S^2 \times S^1$ look like. How could people living in these spaces distinguish them?

14. Prove that the interior of a disk is homeomorphic to $\mathbb{R}^2$.

15. Let $A = \{(x, y) \mid 1 \leqslant x^2 + y^2 \leqslant 2\}$ and let $B = \{(x, y, z) \mid x^2 + y^2 = 1$ and $0 \leqslant z \leqslant 1\}$. Draw pictures of the sets $A$ and $B$ and prove that they are homeomorphic.

16. Let $K_1$ and $K_2$ be knots in $\mathbb{R}^3$. Prove that there exists a homeomorphism $h : (\mathbb{R}^3, K_1) \to (\mathbb{R}^3, K_2)$ if and only if there exists a homeomorphism $g : (S^3, K_1) \to (S^3, K_2)$.

17. Let $K_1$ be the right-handed trefoil knot and let $K_2$ be the left-handed trefoil knot. Is there a homeomorphism $h : (\mathbb{R}^3, K_1) \to (\mathbb{R}^3, K_2)$? Explain.

18. Prove that the composition of two homeomorphisms is a homeomorphism.

19. Suppose $f_1$, $f_2$, $g_1$, and $g_2$ are all homeomorphisms of $S^3$. Suppose that $f_1$ is isotopic to $g_1$ and $f_2$ is isotopic to $g_2$. Prove that $f_2 \circ f_1$ is isotopic to $g_2 \circ g_1$.

20. Explain why a mirror reverses your left and your right sides but not your head and your feet.

21. A relation $\sim$ on a set $S$ is said to be an *equivalence relation* if, for every $a, b, c \in S$ we have: (1) $a \sim a$; (2) if $a \sim b$ then $b \sim a$; (3) if $a \sim b$ and $b \sim c$, then $a \sim c$. Prove that knot equivalence is an equivalence relation.

22. Prove that one of the kink moves that is illustrated in Figure 2.3 can be derived from the other one by using a move from Figure 2.4 or Figure 2.5.

23. Compute the P-polynomial of the figure eight knot, which is illustrated below.

24. Use the P-polynomial to prove that the unoriented (6, 2)-torus link is topologically chiral.

25. Can we use Kauffman's method to distinguish the embedded rigid vertex graphs in Figure 2.17? Explain.

26. Prove that a pair of links $L_1$ and $L_2$ are equivalent if and only if there is an orientation-preserving homeomorphism $h : \mathbb{R}^3 \to \mathbb{R}^3$ such that $h(L_1) = L_2$.

27. How many copies of a trefoil knot are contained in $T(G)$ if $G$ is the graph of the molecular trefoil knot?

28. Draw a nonalternating diagram of a nontrivial knot, which is different from those in the text.

29. Prove that the knot in Figure 2.33 is topologically chiral.

30. Explain why two components of a link must always cross an even number of times in a link projection.

31. Prove that the linking number is a topological invariant.

32. Find deformations transforming the link projection of $9^2_{61}$ in Figure 2.34 to that of Figure 2.37, and the projection of Figure 2.38 to the one of Figure 2.39.

33. Prove that we can go from the projection of a figure eight knot to the mirror image projection by using a sequence of Reidemeister moves 0, 2, and 3, without having to use move number 1.

34. Consider the triple layered cyclophane illustrated below. Determine whether this structure is topologically chiral, chemically chiral, whether it is a topological rubber glove, and whether it is a Euclidean rubber glove. Justify each of your conclusions.

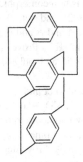

35. Find a ten-crossing projection of the following knot.

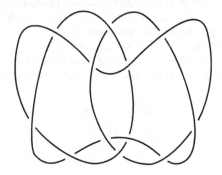

36. Let $f : (0, 1) \to (1, \infty)$ given by $f(x) = \frac{1}{x}$. Prove that $f$ is a homeomorphism.

37. Suppose that $f$ and $g$ are both homeomorphisms of $\mathbb{R}^3$. Prove that if $f$ is isotopic to $g$, then $g$ is isotopic to $f$.

38. Suppose that a link with components $J$ and $K$ is topologically achiral as an oriented link (i.e. it can be deformed to its mirror image preserving the orientation on each component). Prove that $\text{Lk}(J, K) = 0$.

39. Find a link with components $J$ and $K$ which is topologically achiral as an unoriented link and $\text{Lk}(J, K)$ is non-zero.

40. Find a link with components $J$ and $K$ which is topologically chiral as an oriented link yet has $\text{Lk}(J, K) = 0$. Prove that your link is topologically chiral.

# 3

# Möbius Ladders and Related
# Molecular Graphs

The molecular bond graph of a three-rung Möbius ladder was the first molecular graph that required sophisticated topological machinery to prove that it could not be deformed to its mirror image (see Figure 3.1). Because of its form as a Möbius strip, this molecule appeared to be distinct from its mirror image. There was also experimental evidence indicating that the molecule was chemically chiral (Walba, 1983). However, because the molecule was a ladder rather than a strip and it was both long and flexible, in theory it might be able to deform itself to its mirror image in some nonobvious way. Walba conjectured that the molecular bond graph of the Möbius ladder was, in fact, topologically chiral, although he could not prove it. In this chapter we shall explain Jon Simon's proof (Simon, 1986) that the embedded graphs representing the molecular Möbius ladders with three or more rungs are topologically chiral. Then we will prove that every embedding of a Möbius ladder with an odd number of rungs greater than two is necessarily topologically chiral. In contrast, recall from Chapter 1 that every Möbius ladder with an even number of rungs has a topologically achiral embedding. Finally, we will discuss the topological chirality of some molecular graphs that are related to Möbius ladders.

## Chirality of Möbius Ladders

In order to consider the general form of the graph of a Möbius ladder, we would like to omit all of the vertices of valence 2. However, the vertices of valence 2 enabled us to distinguish easily between those edges representing the sides of the ladder and those edges representing the rungs of the ladder. Walba (1983) observed that if we consider the standardly embedded three-rung Möbius ladder (that is the one that is embedded in the form of a Möbius strip), without differentiating the edges that represent rungs from the edges that represent sides, then this embedded graph is topologically achiral. In particular, if we ignore the

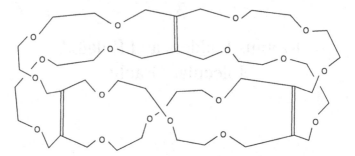

**Figure 3.1.** A molecular Möbius ladder

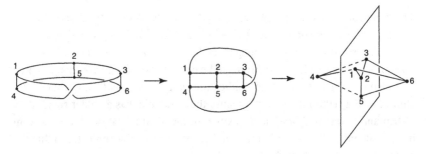

**Figure 3.2.** If we do not distinguish the two types of edges, a Möbius ladder with three rungs can be deformed to a symmetry presentation with a plane of reflection

difference between these two types of edges, then the three-rung Möbius ladder can be deformed to a symmetry presentation that has a planar reflection. This deformation is illustrated in Figure 3.2. The first step illustrated in the picture is that we pull the edge connecting vertices 3 and 4 down, and we pull the edge connecting vertices 1 and 6 up to get the second picture. Next we deform the edges connecting vertices 1, 2, 3, and 5 into the shape of a Y in a vertical plane, with vertices 4 and 6 on opposite sides of the plane. This vertical plane is the mirror plane that reflects the Möbius ladder back to itself. The planar reflection that is illustrated in Figure 3.2 shows that the three-rung Möbius is topologically achiral. However, observe that this planar reflection interchanges some of the sides of the ladder with some of the rungs. For example, in Figure 3.2 the rung from vertex 1 to vertex 4 is interchanged with the side from vertex 1 to vertex 6.

The planar reflection in Figure 3.2 cannot exist for the actual graph of the molecular Möbius ladder, because the edges that are rungs are carbon–carbon double bonds, whereas the edges on the sides of the ladder are made up of

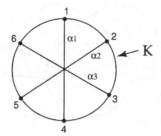

**Figure 3.3.** An abstract representation of a Möbius ladder with three rungs

single bonds that are contained in the chain of oxygens and carbons. So from a chemical point of view, an edge that is a rung cannot be exchanged with an edge that is a side of the ladder. We shall formally rephrase Walba's conjecture about the topological chirality of molecular Möbius ladders while making the distinction between the different types of edges, after we define the abstract graph of a Möbius ladder.

**Definition.** A *Möbius ladder*, $M_n$, is the graph consisting of a simple closed curve $K$ with $2n$ vertices, together with $n$ additional edges $\alpha_1, \ldots, \alpha_n$ such that if the vertices on $K$ are consecutively labeled $1, 2, 3, \ldots, 2n$ then the vertices of each edge $\alpha_i$ are $i$ and $i + n$. We say $K$ is the *loop* of $M_n$ and $\alpha_1, \ldots, \alpha_n$ are the *rungs* of $M_n$.

Figure 3.3 is an illustration of the abstract graph $M_3$. The edges $\alpha_1, \alpha_2$, and $\alpha_3$ are not in fact permitted to intersect in an embedding of the graph. We draw them as intersecting rather than specifying a particular embedding. The definition of $M_n$ does not actually depend on the labeling of the vertices. This labeling is just a convenient way to make it clear that the rungs of the Möbius ladder join vertices that are halfway around on the simple closed curve $K$. Notice that we can choose the loop of $M_3$ in several different ways. For example, there is the obvious loop, which we have labeled $K$ in Figure 3.3, which has vertices consecutively numbered 1, 2, 3, 4, 5, and 6. However, we can also think of $M_3$ as a Möbius ladder, where the loop contains the edges $\alpha_1$, $\alpha_2$, and $\alpha_3$ and traverses the vertices in the order 1, 2, 5, 6, 3, and 4 or in the order 1, 6, 3, 2, 5, and 4.

We remark here that no matter how the loop $K$ of a Möbius ladder $M_n$ is chosen, cutting $K$ apart at the vertices of any rung $\alpha$ will split $K$ into two components that each contain $n$ edges. This is an important property of the relationship between the loop and the rungs of a Möbius ladder.

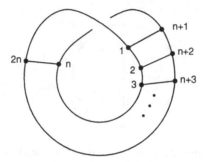

**Figure 3.4.** A standard embedding of $M_n$

We say that $M_n$ is *standardly embedded* in $S^3$, if $M_n$ is embedded as in Figure 3.4 or its mirror image. As explained above, for chemical reasons we shall distinguish between the rungs and the sides of the graph, so we want any homeomorphism of $M_n$ to take rungs to rungs and sides to sides. In particular, any such homeomorphism will take each edge contained in the loop $K$ to an edge that is also contained in $K$. Thus the loop $K$ will be preserved by such a homeomorphism. Conversely, if the loop $K$ is preserved by $h$, then $h$ will take rungs to rungs and sides to sides. We rephrase Walba's question as asking whether, for a standardly embedded $M_n$, there exists an orientation-reversing homeomorphism $h : (\mathbb{R}^3, M_n) \rightarrow (\mathbb{R}^3, M_n)$ such that $h(K) = K$. Recall from Chapter 1 (Figure 1.7) that there is such a homeomorphism if $n = 2$, and clearly there is also such a homeomorphism if $n = 1$. Jon Simon (1986) answered Walba's question in the negative for any $n \geqslant 3$. Furthermore, he proved that the assumption that $h(K) = K$ is not necessary if $n \geqslant 4$. In particular, for $n \geqslant 4$, he proved that every automorphism of $M_n$ sends the sides of the ladder to the sides and the rungs of the ladder to the rungs (recall that an automorphism of a graph is a one-to-one and onto map, taking vertices to vertices in such a way that adjacent vertices are taken to adjacent vertices). In other words, for $n \geqslant 4$, there is only one way to choose the loop of $M_n$. Thus, for $n \geqslant 4$, no matter how $M_n$ is embedded in $\mathbb{R}^3$, if there exists a homeomorphism $h : (\mathbb{R}^3, M_n) \rightarrow (\mathbb{R}^3, M_n)$ then it follows that $h(K) = K$. We will prove Simon's result about automorphisms of $M_n$ first (Simon, 1986), and then we will prove the topological chirality of the standardly embedded Möbius ladders.

**Lemma 3.1.** *For* n $\geqslant 4$, *every automorphism of the vertices of the graph* $M_n$ *leaves the loop* K *setwise invariant.*

*Proof.* Suppose that $A$ is an automorphism of $M_n$ that does not leave $K$ setwise invariant. Then $A$ must send $K$ to some other simple closed curve $H$ with $2n$

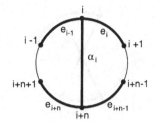

**Figure 3.5.** The edges connected to $\alpha_i$

edges, such that $M_n$ can also be considered as a Möbius ladder with loop $H$. Because $H \neq K$, there is some edge $\alpha_i$ that is contained in $H$ but is not contained in $K$. Thus, if we consider $M_n$ with loop $K$ and vertices labeled consecutively going around $K$, then $\alpha_i$ is a rung that we can assume has vertices $i$ and $i + n$. Because $H$ is a simple closed curve containing $\alpha_i$, there must be precisely one other edge of $H$ with vertex $i$. The edge $e_{i-1}$ with vertices $i - 1$ and $i$, and the edge $e_i$ with vertices $i$ and $i + 1$ are the only other edges of $M_n$ containing the vertex $i$. We illustrate these edges for the abstract graph $M_n$ in Figure 3.5. In order to focus attention on $\alpha_i$, we do not include the other rungs of $M_n$.

Without loss of generality, we shall assume that $H$ contains the edge $e_{i-1}$, but not the edge $e_i$. Also, because $\alpha_i$ contains the vertex $i + n$, the loop $H$ must contain precisely one other edge with the vertex $i + n$. Thus $H$ contains either the edge $e_{i+n-1}$ with vertices $i + n - 1$ and $i + n$, or the edge $e_{i+n}$ with vertices $i + n$ and $i + n + 1$. We consider these cases separately.

*Case 1.* $H$ contains the edge $e_{i+n-1}$.

By hypothesis, the loop $H$ contains the edge $e_{i-1}$ with vertices $i - 1$ and $i$, the edge $\alpha_i$ with vertices $i$ and $i + n$, and the edge $e_{i+n-1}$ with vertices $i + n$ and $i + n - 1$. If $H$ also contained the edge $\alpha_{i-1}$ of $M_n$ that has vertices $i - 1$ and $i + n - 1$, then $H$ would contain a simple closed curve made of the four edges $e_{i-1}, \alpha_i, e_{i+n-1}$, and $\alpha_{i-1}$. This simple closed curve is illustrated with darkened lines in Figure 3.6.

This is impossible because $H$ is a simple closed curve containing $2n$ edges, and $n \geqslant 4$. Thus, $H$ does not contain the edge $\alpha_{i-1}$. So $\alpha_{i-1}$ must be a rung when $M_n$ is considered to have loop $H$. As we remarked after our definition of a Möbius ladder, if $H$ is cut open at the vertices of a rung, then each of the two components must contain $n$ edges. We have seen above that the edges $e_{i-1}$, $\alpha_i, e_{i+n-1}$, and $\alpha_{i-1}$ form a simple closed curve. Hence one component of $H$ after the removal of the vertices of the rung $\alpha_{i-1}$ contains only the edges $e_{i-1}$, $\alpha_i$, and $e_{i+n-1}$. This is impossible since $n \geqslant 4$. Therefore, Case 1 cannot arise.

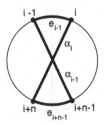

**Figure 3.6.** A simple closed curve made of the edges $e_{i-1}$, $\alpha_i$, $e_{i+n-1}$, and $\alpha_{i-1}$

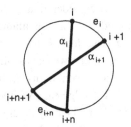

**Figure 3.7.** The edges $\alpha_i$, $e_{i+n}$ and $\alpha_{i+1}$

*Case 2.* $H$ contains the edge $e_{i+n}$.

The loop $H$ must contain every vertex of $M_n$. So, in particular, $H$ contains the vertex $i+1$. Since $H$ is a simple closed curve, $H$ must include two edges which contain each vertex. We know by our initial assumption that $H$ contains $e_{i-1}$ but does not contain $e_i$. Thus, in particular, $H$ must contain $\alpha_{i+1}$. The edges $\alpha_i$, $e_{i+n}$ and $\alpha_{i+1}$ in $H$ are illustrated as darkened edges in Figure 3.7.

Because $H$ does not contain $e_i$, it must be that $e_i$ is a rung when $M_n$ is considered to have loop $H$. So by the remark after our definition of a Möbius ladder, if we remove the vertices of the rung $e_i$, then $H$ will be split into two components that each contain $n$ edges. However, one component of $H - e_i$ consists of the union of the edges $\alpha_i$, $e_{i+n}$, and $\alpha_{i+1}$, which is again impossible as $n \geq 4$. Thus this case does not occur either.

Hence no such automorphism exists.                              □

In order to understand Simon's proof that standardly embedded Möbius ladders are topologically chiral, as well as subsequent proofs, the reader needs to have a good understanding of twofold branched covers. Such covers are also an important tool in the study of knot theory and three-dimensional topology, so we now take a fairly long detour to discuss twofold branched covers. We begin with a couple of preliminary definitions for readers who do not have a strong background in topology. A set $V$ in Euclidean $p$-dimensional space, $\mathbb{R}^p$, is said to be *open* if for every point $x \in V$, there is some $\varepsilon > 0$ such that if $\|x - y\| < \varepsilon$ then $y \in V$. This means that for every point $x$ of $V$ there is some $\varepsilon > 0$ such that $x$ is contained in the interior of a ball of radius $\varepsilon$ that is a subset of $V$.

**Definition.** Let $M$ be a subset of $\mathbb{R}^p$ for some $p$. A subset $U$ of $M$ is said to be *open* in $M$ if $U = M \cap V$ where $V$ is an open set in $\mathbb{R}^p$. Let $n$ be a natural number. We say that $M$ is an *n-manifold* if each point $x$ of $M$ is contained in an open set $U$ of $M$ that is either homeomorphic to $\mathbb{R}^n$ or to the half-space $\mathbb{R}^n_+ = \{(x_1, \ldots, x_n) \in \mathbb{R}^n | x_n \geq 0\}$.

Note that $n$ does not necessarily equal $p$. Intuitively, an $n$-manifold is a space that locally looks like $\mathbb{R}^n$ or $\mathbb{R}^n_+$. For example, the surface of a doughnut (which we call a torus) is a two-manifold, while a solid doughnut (which we call a solid torus) is a three-manifold. Both of these manifolds are subsets of $\mathbb{R}^3$. We can see as follows that a torus satisfies our definition of a two-manifold. We can imagine a solid torus embedded in $\mathbb{R}^3$ as an ordinary doughnut, and let $T$ denote the surface of the doughnut. For any point $x \in T$, we can choose $V$ to be a small open ball in $\mathbb{R}^3$, which is centered at $x$. Because $V$ is open in $\mathbb{R}^3$, the set $U = T \cap V$ is open in $T$ by definition. When $V$ is chosen small enough, $U$ will be the interior of a small curved disk in the surface $T$. The interior of a curved open disk is homeomorphic to the interior of a flat disk, which in turn is homeomorphic to $\mathbb{R}^2$. Every point in $T$ is contained in the interior of such a disk, so $T$ is a two-manifold. We can similarly argue that a solid torus $S$ is a three-manifold. In this case, the points in the interior of $S$ are each contained in an open set that is homeomorphic to $\mathbb{R}^3$, whereas the points in the boundary of $S$ are each contained in an open set that is homeomorphic to $\mathbb{R}^3_+$. Thus a solid torus is a three-manifold with boundary. We can also have a two-manifold with boundary. For example, a Möbius strip and an annulus (i.e., a disk with a hole cut out of it) are both two-manifolds with boundary. We often refer to two-manifolds as *surfaces*. The three-dimensional sphere $S^3 = \mathbb{R}^3 \cup \{\infty\}$, $\mathbb{R}^3$, and the three-dimensional ball $B^3$ are all examples of three-manifolds. A circle, a line, a line segment, and unions of these are the only one-manifolds. There is a well-known classification of all two-manifolds (see, e.g., Massey, 1991), but there is no known classification of all three-manifolds. An arbitrary graph is not a one-manifold, because at any vertex with a valence of at least three, there is no open set that is homeomorphic to $\mathbb{R}^1$ or $\mathbb{R}^1_+$. In general, we are interested in manifolds because they are nicer than other subsets of $\mathbb{R}^n$. Graphs will be the only sets that we are interested in that are not manifolds.

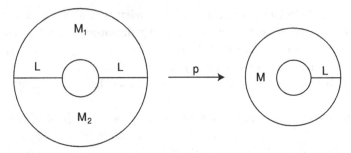

**Figure 3.8.** The map $p$ sends each of $M_1$ and $M_2$ onto the set $M$

**Definition.** Let $M$ be a subset of $\mathbb{R}^n$, and let $h : M \rightarrow M$ be a homeomorphism. Let $r$ be a natural number. We define $h^r$ as the homeomorphism obtained by performing the map $h$ repeatedly $r$ times. If $r$ is the smallest number such that $h^r$ is the identity map, then we say that $h$ has an *order* equal to $r$. If there is no such number $r$, then we say that $h$ does not have a finite order.

Any rigid motion that takes a graph to itself in $\mathbb{R}^3$ has some finite order $r$. For instance, any reflection has order two, whereas a rotation of the graph by $120°$ about an axis has order three.

In order to motivate our definition of twofold covers and branched covers of three-manifolds, we first consider some two-dimensional examples. Let $M$ represent the annulus obtained by removing the interior of a unit disk from a disk of radius two. We can express the points of $M$ in polar coordinates as $(r, \theta)$ where $1 \leqslant r \leqslant 2$. We shall define an order-two homeomorphism $h : M \rightarrow M$ by $h(r, \theta) = (r, \theta + 180°)$. Then $h$ rotates $M$ by $180°$, so if we perform $h$ twice, we will obtain the identity map. Now define a function $p : M \rightarrow M$ by $p(r, \theta) = (r, 2\theta)$. The functions $h$ and $p$ are related because for every $x, y \in M$, we have $p(x) = p(y)$ if and only if either $x = y$ or $h(x) = y$.

We can think of $p$ as wrapping the annulus $M$ twice around itself. We shall let $M_1$ be the surface obtained by cutting $M$ open along a line segment of the form $L = \{(r, 0) | 1 \leqslant r \leqslant 2\}$. We can stretch $M_1$ open so that it looks like half of an annulus, and we can let $M_2$ be a copy of $M_1$. Then we obtain $M$ again if we glue $M_1$ to $M_2$ along the two line segments that came from the cut $L$. Observe that the map $h$ interchanges the sets $M_1$ and $M_2$, and the map $p$ sends each of $M_1$ and $M_2$ onto the set $M$ (see Figure 3.8). In this example, $M$ is the twofold cover of itself.

Now we consider a slightly harder example. Let $M$ denote a disk with three holes cut out of it, and let $N$ denote a disk with two holes cut out of it, as illustrated in Figure 3.9. Let $h : M \rightarrow M$ be a rotation by $180°$ about a point in the center of the middle hole. Let $M_1$ be the surface obtained by cutting $N$ along

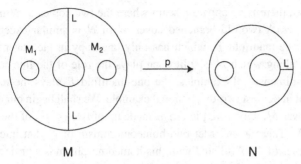

**Figure 3.9.** The map $p$ sends each of $M_1$ and $M_2$ onto $N$ by gluing together the two copies of $L$

the line segment indicated in Figure 3.9, and let $M_2$ be a copy of $M_1$. If we stretch $M_1$ and $M_2$ open and glue them together as illustrated on the left in Figure 3.9, then we will obtain $M$. We define $p : M \to N$ to be the map that sends each of $M_1$ and $M_2$ onto $N$ by gluing together the two copies of $L$. Observe that the map $h$ interchanges $M_1$ and $M_2$, and $p(x) = p(y)$ if and only if either $x = y$ or $h(x) = y$. This is an example of where $M$ is a twofold cover of $N$.

Our final two-dimensional example is similar to our first example; however, it uses a disk instead of an annulus. We let $M$ denote a unit disk expressed in polar coordinates. We define $h : M \to M$ by $h(r, \theta) = (r, \theta + 180°)$, so that $h$ rotates $M$ by $180°$. Now define $p : M \to M$ by $p(r, \theta) = (r, 2\theta)$. As in our previous examples, $p(x) = p(y)$ if and only if either $x = y$ or $h(x) = y$. We let $M_1$ be the surface obtained by cutting $M$ open along a single radius of the disk, and we let $M_2$ be a copy of $M_1$. By stretching $M_1$ and $M_2$ open, we can think of each as a half-disk, and the disk $M$ is obtained by gluing these two half-disks together. Thus again $h$ interchanges $M_1$ and $M_2$, and $p$ sends each of $M_1$ and $M_2$ onto $M$. What makes this example different from the previous two examples is that in the first two examples $h$ fixed no points of $M$, whereas in this example $h$ fixes the center point of $M$. This is an example where $M$ is a twofold branched cover of itself with branch set the center point of $M$.

These examples of surfaces that are twofold covers and branched covers help to motivate the definition of twofold covers and branched covers of three-manifolds. Because we are interested only in twofold coverings of three-manifolds, rather than arbitrary covering spaces, we will give a specialized definition. For a more general treatment of covering spaces, see Munkres (1975). Intuitively, a twofold cover $M$ of a three-manifold $N$ consists of two copies of the set obtained by cutting $N$ open along a surface and then gluing together these two copies in such a way that, locally, $M$ looks exactly the same as $N$.

In particular, there is no apparent seam where the two copies of $N$ have been sewn together. A twofold branched cover $M$ of $N$ is similar except that $N$ contains a one-manifold $B$, which has only one copy in the cover $M$. Thus the surface along which $N$ is split open plays the role of the line segment $L$ in our two-dimensional examples; the one-manifold $B$ plays the role of the center point in our last two-dimensional example. We shall begin our definition with the cover $M$, which we picture as made up of two copies of the cut open manifold $N$. There is an order-two homeomorphism of $N$ that interchanges the two copies of $M$. Affiliated with this homeomorphism is a projection map $p : N \rightarrow M$ taking each of the copies of the cut open $N$ that live in $M$ onto the real $N$. Formally, we have the following definition.

**Definition.** Let $M$ and $N$ be three-manifolds, and let $h : M \rightarrow M$ be an orientation-preserving homeomorphism of order two. Let $p : M \rightarrow N$ be a function such that $p(x) = p(y)$ if and only if either $x = y$ or $h(x) = y$. Suppose that $p$ is a continuous onto map that takes open sets to open sets. Let $A$ denote the set of points $x$ in $M$ such that $h(x) = x$. If $B = p(A)$ is a one-manifold, then we say that $M$ is a *twofold branched cover* of $N$ branched over $B$. If $A$ is the empty set then we say that $M$ is a *twofold cover* of $N$. In either of these cases, we say that $p$ is the *projection map*, $N$ is the *base space*, and $h$ is the *covering involution*.

Those with a background in topology will also observe that standard covering space theory guarantees that when the base space $N$ has either zero or one boundary component and the boundary of $B$ meets each component of the boundary of $N$ in an even number of points, then the twofold branched cover of $N$ always exists. Furthermore, if the first homology group $H_1(N; \mathbb{Z}_2) = 0$, then $M$ is unique in the sense that if $M'$ is another twofold branched cover of $N$ branched over $B$ with projection map $p' : M' \rightarrow N$, then there exists a homeomorphism $\varphi : M \rightarrow M'$ such that $p = p' \circ \varphi$. We shall be considering twofold branched covers of $S^3$ in this chapter and twofold branched covers of the three-dimensional ball, $B^3$, in Chapter 7. The uniqueness of the twofold branched covers of $S^3$ and of the three-dimensional ball branched over any specified one-manifold follows from the above comment. Because of this uniqueness, we shall speak of *the* twofold branched cover of $S^3$ branched over a simple closed curve $K$, rather than *a* twofold branched cover of $S^3$ branched over a simple closed curve $K$.

The definition of a twofold branched cover may seem very complicated. However, it corresponds to the examples we gave for surfaces, and we will shortly see some examples of twofold branched covers of three-manifolds.

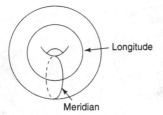

**Figure 3.10.** The meridian and longitude of a solid torus

We will start with $N = S^3$ together with a simple closed curve $K$, which will be our branch set. Then we will simultaneously construct $M$ and the maps $h$ and $p$. Note that for a given $N$ we can have different $M$s, depending on what $K$ is. To illustrate our definition, we give a couple of detailed examples. The first example that we shall construct is the twofold branched cover of $S^3$ branched over an unknot $K$.

Both of our examples of twofold branched covers will rely on an understanding of solid tori in $S^3$. Figure 3.10 illustrates two important curves on the surface of a solid torus embedded in $S^3$. Formally, a *meridian* of a solid torus is defined as a simple closed curve on the boundary of the solid torus that bounds a disk in the interior of the solid torus but does not bound a disk in the boundary of the solid torus. A disk whose boundary is a meridian is called a *meridional disk*. A *longitude* of a solid torus embedded in $S^3$ is defined as a simple closed curve on the boundary of the solid torus that bounds a two-sided surface in the complement of the solid torus but does not bound a disk on the boundary of the solid torus. For an unknotted solid torus, like the one in Figure 3.10, a longitude actually bounds a disk in the complement of the solid torus. However, if the solid torus is knotted in $S^3$, then a longitude bounds a more complex surface. We shall see how to construct such a surface in our second example of a twofold branched cover. A *meridional rotation* of the solid torus is one that sends each meridional disk to itself, fixing its center point, and sends each longitude to a disjoint longitude. In contrast, a *longitudinal rotation* sends each longitude to itself and each meridional disk to a disjoint meridional disk.

To build our example, we shall use the fact that $S^3$ can be constructed as the union of two unknotted solid tori, glued together along their boundaries in such a way that the longitude of one is glued to the meridian of the other. Figure 3.10 illustrates one of the two solid tori, and the other solid torus is the complement of the interior of this solid torus in $S^3$. You can imagine a tube coming up through the center of the solid torus in Figure 3.10, wrapping around the outside and closing up at the point of $\infty$. The curve that is labeled "longitude" is also

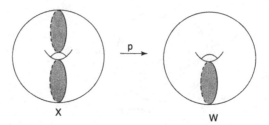

**Figure 3.11.** The map $p$ longitudinally wraps the first torus twice around the second torus, sending two meridional disks of $X$ to one meridional disk of $W$

the meridian of the complementary solid torus, and the curve that is labeled "meridian" is also the longitude of the complementary solid torus.

Now we begin the construction of our twofold branched cover. Let $V$ be a solid torus with core the unknotted circle $K$. That is, $V$ is a small tube encasing $K$, where $K$ runs along the central core of $V$. If $K$ is a round circle then $V$ looks like the solid torus in Figure 3.10, whereas, if $K$ wiggles then $V$ will wiggle in a corresponding way. We shall let $W$ denote the solid torus that is the complement in $S^3$ of the interior of the unknotted solid torus $V$. The circle $K$ is going to be the branch set of our twofold branched cover. Since $K$ is the core of $V$, the twofold branched cover of $S^3$ branched over $K$ can be obtained by appropriately gluing together the twofold cover of $W$ and the twofold branched cover of $V$ branched over $K$.

We shall see as follows that the twofold cover of the solid torus $W$ is just another solid torus $X$. In particular, we cut $W$ open along a meridional disk and stretch it open so that it looks like a half-doughnut. On each end of the half-doughnut is a meridional disk. Now $X$ consists of two copies of this half-doughnut glued together along the two meridional disks to obtain a complete doughnut. We let the covering involution be a 180° longitudinal rotation of the solid torus, which sends each meridional disk to another meridional disk that is halfway around the solid torus. Now the projection map $p : X \to W$ is a two-to-one map that wraps each of the halves of $X$ onto $W$, and that glues together the two meridional disks that were in the boundary of each half of $X$. In Figure 3.11 we have indicated two meridional disks of $X$ that both get mapped by $p$ to the same meridional disk of $W$. We can see intuitively that $p$ is a continuous onto map that takes open sets to open sets. Thus $X$ is the twofold cover of $W$.

The twofold branched cover of the solid torus $V$ branched over its core $K$ turns out also to be a solid torus, which we will call $U$. To see this, we cut $V$ open along an annulus that is contained inside the solid torus and that has one boundary as a longitude on the boundary torus and the other boundary as the

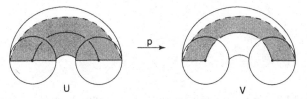

**Figure 3.12.** The projection map $p$ wraps meridional disks of $U$ twice around the corresponding meridional disks of $V$

core of the solid torus. This annulus is shaded in the half-solid torus illustrated as $V$ in Figure 3.12. After cutting $V$ open along this annulus we can stretch it open and flatten it out so that it looks like the bottom half of a bagel that has been cut in half to make a sandwich. Gluing this bottom half of a solid torus to an identical top half of a solid torus (as we would if we were to make a bagel sandwich), we get the complete solid torus $U$. We let the covering involution $h$ of $U$ be a 180° meridional rotation, which leaves each meridional disk setwise invariant, while fixing the point in the center of each meridional disk. Then $h$ interchanges the top and the bottom half of the bagel and sends each longitude to a disjoint longitude that is halfway around the torus. In particular, $h$ sends the longitude that goes around the outside of the bagel to the one that goes around the hole of the bagel. Every point on the central core of $U$ is fixed by the covering involution $h$. The projection map $p : U \to V$ sends both the top and the bottom half of $U$ onto $V$ by wrapping each half of a meridional disk of U onto an entire meridional disk of $V$. In Figure 3.12 we illustrate a cross section of the solid tori $U$ and $V$ where two radii of each meridional disk of $U$ get sent to a single radius of a meridional disk of $V$, and the shaded annulus in $U$ gets folded along its central circle and sent to the shaded annulus in $V$.

Now we must glue together the solid tori $X$ and $U$ in a way that corresponds to the way that $W$ was originally glued to $V$. That is, in order for $p$ to be well defined on the intersection of $X$ and $U$, we must glue $X$ and $U$ together in such a way that if a point $a \in X$ is glued to a point $b \in U$ then $p(a) \in W$ is glued to $p(b) \in V$. Furthermore, we must glue the solid torus $X$ to the solid torus $U$ in such a way that the longitudinal rotation of the covering involution on the boundary of $X$ matches up with the meridional rotation of the covering involution on the boundary of $U$. In this way, both the projection map and the covering involution will be well defined on the torus that is the intersection of $X$ and $U$. It follows that each meridian of one solid torus must be glued to a longitude of the other solid torus. By gluing the solid tori together in this way, we end up with $S^3$ again. Thus $S^3$ is the twofold branched cover of $S^3$ branched over the unknot.

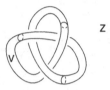

**Figure 3.13.** $Z$ is the complement of the interior of the knotted solid torus $V$

For our next example we want to construct the twofold branched cover of $S^3$ branched over a nontrivial knot $K$. Let $V$ be a solid torus with core the knot $K$. So, $V$ is a knotted tube corresponding to the way that $K$ is knotted in $S^3$. Comparing this example with the last one, we see that $V$ is again a solid torus, but what is different now is that the complement of the interior of $V$ is no longer a solid torus. Rather, it is $S^3$ with a knotted hole cut out. Because $V$ is itself a solid torus, the twofold branched cover of $V$ branched over $K$ is the solid torus $U$ that we obtained above, where the covering involution on $U$ is again a 180° meridional rotation. Now let $Z$ denote the complement of the interior of the knotted solid torus $V$. We call any space that is obtained this way a *knot complement*. We illustrate a possible $V$ and $Z$ in Figure 3.13.

In our last example, we constructed the twofold cover of a solid torus by first cutting the solid torus open along a disk and then gluing two copies of this half-doughnut together. To find the twofold cover of $Z$, we first need to construct a surface bounded by the knot $K$ that we will cut $Z$ open along. However, we do not want just any surface bounded by the knot. Rather, we want a surface that has two distinct sides, so that when we cut $Z$ open, one side of the surface will be adjacent to the interior of $Z$ and the other side will be adjacent to the exterior of $Z$.

Two-sided surfaces are related to the concept of orientability, which we explain as follows. An *orientation-reversing path* in an $n$-manifold is a simple closed curve that has the property that if an $n$-dimensional object is moved once around this curve, then it would return to its original position as its mirror image. For example, if a two-manifold contains a Möbius strip, then a simple closed curve that goes once around this Möbius strip is an orientation-reversing path because a small circle in the surface that is oriented clockwise would return to its original position as a circle that is oriented counterclockwise. We can see this in Figure 3.14, where we begin with a clockwise circle on the far right of the Möbius strip; after it travels all the way around, it returns as a counterclockwise circle. An orientation-reversing path in a three-manifold is a simple closed curve, which you could walk along and then return to your original position with your left and your right sides reversed. In particular, when you returned your heart would be on the right side of your body.

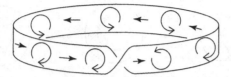

**Figure 3.14.** A clockwise circle becomes a counterclockwise circle when it travels around a Möbius strip

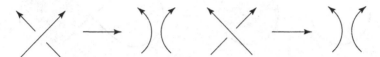

**Figure 3.15.** Each crossing is replaced by a pair of parallel strands

An $n$-manifold is said to be *orientable* if it does not contain an orientation-reversing path. In particular, an orientable surface is one that does not contain a Möbius strip. Observe that in $\mathbb{R}^3$, a disk or a torus has two sides whereas a Möbius strip has only one side. It is not difficult to prove that for surfaces embedded in an orientable three-manifold (for instance, the three-manifolds $S^3$ or $\mathbb{R}^3$), the orientable surfaces are precisely those that have two distinct sides (see Exercise 11).

We show as follows how to construct an orientable surface that is bounded by a knot in $S^3$. First we arbitrarily orient a projection of the knot $K$. Then we temporarily eliminate the crossings of $K$ by replacing them by parallel strands, as indicated in the diagrams in Figure 3.15. This operation is called *smoothing* the crossings of a knot.

Eliminating the crossings in this way gives us a collection of disjoint (possibly nested) circles in the plane. We think of these circles as the boundaries of disjoint disks, where any nested disks are considered to be at different heights above the page. Now we connect these disks with half-twisted bands in such a way that the projections of the half-twisted bands are the same as the original crossings of the knot. The surface that we get, in this way, has $K$ as its boundary and is called a *Seifert surface* for $K$. We illustrate in Figure 3.16 how to go from an oriented trefoil knot to stacked disks, and finally to a Seifert surface for the knot. Here the Seifert surface consists of two disks joined together by three half-twisted bands.

We shall see below that the Seifert surface that we obtain in this way is always two-sided. Observe that the orientation of the knot induces an orientation on each of the disjoint circles that we obtained by smoothing all of the crossings.

**Figure 3.16.** We construct the Seifert surface for a trefoil knot

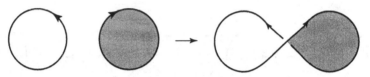

**Figure 3.17.** If two adjacent circles had a crossing between then, then we can add a half-twisted band to obtain a two-sided surface

If the orientation of the boundary of a disk is clockwise from our viewpoint, then we say that the side of the disk that is facing us is the *negative side*; if the orientation of the boundary of a disk is counterclockwise, then we say that the side of the disk that is facing us is the *positive side*. The underside of each disk has the opposite sign. Thus in Figure 3.16, both disks have their positive sides facing up and their negative sides facing down.

When we smooth a crossing we obtain two little parallel arrows. So if two adjacent (that is, nonnested) circles originally had a crossing between them, then the pair of circles must have opposite orientations, one clockwise and the other counterclockwise. Thus one disk has its positive side up and the other disk has its negative side up. Hence we can assign positive and negative sides to the half-twisted band connecting these disks in such a way that the sign associated with each side of the half-twisted band agrees with each disk along the arc where the band is attached. Therefore, by connecting the band to the two disks, we still have a two-sided surface. We illustrate this process in Figure 3.17 by coloring the negative sides gray, and the positive sides white.

In contrast, if two nested circles originally had a crossing between them, then the pair of circles must have the same orientation (as was the case in Figure 3.16), because the two little parallel arrows either both go clockwise or both go counterclockwise. Thus they each have their positive side facing up or they each have their negative side facing up. Because the half-twisted band joins the underside of the lower disk to the top side of the upper disk, we can again assign a negative and positive side to the band so that it agrees with the disk where the band is attached. In Figure 3.18 the light gray indicates the positive side and the dark gray indicates the negative side of the surface.

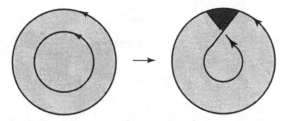

**Figure 3.18.** If two nested circles had a crossing between them, then we can add a half-twisted band to obtain a two-sided surface

By consecutively adding all of the half-twisted bands as above, we see that a Seifert surface can always be assigned a positive and a negative side according to the orientation of the knot, and hence the surface is two-sided. Because our Seifert surface is contained in $S^3$, the Seifert surface is necessarily an orientable surface.

Observe that we can use a similar construction, starting with a link to create a two-sided surface bounded by the link. Although the Seifert surface for a knot does not depend on the orientation of the knot, the Seifert surface for a link may change if we change the orientation of one or more components of the link. In any case, every link bounds an oriented Seifert surface.

Now we shall use the Seifert surface of a knot to construct the twofold cover of the space $Z$. Recall that $Z$ was the complement in $S^3$ of the interior of the knotted solid torus $V$. Let $P$ denote the intersection of $Z$ with a Seifert surface for $K$. So, $P$ consists of the Seifert surface $S$ after a small band running along $K$ has been removed. By definition of a longitude, the boundary of $P$ is a longitude of $V$. We cut $Z$ open along the surface $P$, and we let $Y$ denote this cut open three-manifold with a copy of $P$ at each of the ends where $P$ had been in $Z$. Because $P$ is two-sided, it has a positive side and a negative side in $Z$. Let $P_+$ denote the copy of $P$ with its positive side facing outside of $Y$ and let $P_-$ denote the copy of $P$ with its negative side facing outside of $Y$. The boundary of $Z$ was a torus, and when we cut open a torus along a longitude, we obtain an annulus. Thus the boundary of $Y$ is this annulus together with the surfaces $P_+$ and $P_-$ that are attached on the ends. Figure 3.19 schematically represents $Y$, where $P_+$ and $P_-$ are each drawn as a torus with a disk removed. This is not an accurate picture of $Y$, since the boundaries of $P_+$ and $P_-$ are actually knotted.

In order to construct the twofold cover of $Z$, we need to glue together two copies of the cut open manifold $Y$. Thus we let $Y_1$ and $Y_2$ each be a copy of $Y$. The manifold $Y_1$ contains the surfaces $P_+$ and $P_-$ exactly as $Y$ did. To avoid confusion, in $Y_2$ we rename the corresponding surfaces as $Q_+$ and $Q_-$, respectively. Let $X$ be the three-manifold obtained from $Y_1$ and $Y_2$ by gluing

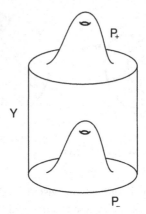

**Figure 3.19.** The boundary of $Y$ is an annulus together with $P_+$ and $P_-$

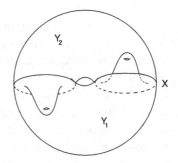

**Figure 3.20.** $X$ is the twofold cover of $Z$

$P_+$ to $Q_-$ and gluing $P_-$ to $Q_+$, matching up pairs of points that came from the same point in $P$. The boundary of $Y_1$ with $P_+$ and $P_-$ removed is an annulus, and the boundary of $Y_2$ with $Q_+$ and $Q_-$ removed is an annulus. So the boundary of $X$ is a torus obtained by gluing the boundary of this annulus in $Y_1$ to the boundary of the corresponding annulus in $Y_2$. The twofold cover of $Z$ will be this manifold $X$, where the covering involution rotates $Y_1$ to $Y_2$, taking $P_+$ to $P_-$. Figure 3.20 is a schematic representation of $X$. Again this picture is not accurate because the boundaries of $P_+$ and $P_-$ should be knotted and $X$ should not be a solid torus.

Recall that $K$ was the core of a solid torus $V$, and the twofold branched cover of $V$ branched along $K$ is a solid torus $U$, where the covering involution is a 180° meridional rotation of $U$. This covering involution sends each longitude of $U$ to a disjoint longitude of $U$. To obtain the twofold branched cover of $S^3$ branched along $K$, we must glue the boundaries of $X$ and $U$ together in such a way that if a point $a \in X$ is glued to a point $b \in U$ then $p(a) \in Z$ is glued to $p(b) \in V$. Furthermore, we must glue the solid torus $U$ to our space $X$ in such a way that the covering involutions on $U$ and $X$ will agree along the torus $U \cap X$. This means that longitudes of $U$ are glued to the boundaries of the surfaces $P_+$ and $P_-$. It follows from our construction that the map $p$ is continuous and onto and takes open sets to open sets. Thus we have constructed the twofold branched cover of $S^3$ branched along $K$. The specific three-manifold that is the twofold branched cover of $S^3$ branched along $K$ depends on what type of knot $K$ is. If $K$ is knotted, then the twofold branched cover of $S^3$ branched over $K$ does not embed in $S^3$, so we cannot draw an accurate picture of the twofold branched cover of $S^3$ branched over a knot. However, we will see that we can still use twofold branched covers as a tool to prove topological chirality of graphs in $S^3$.

We can also generalize this construction to create the twofold branched cover of $S^3$ branched over a link $L$. To do this we first create a collection of solid tori whose cores are the components of the link. We let $Z$ be the complement of the interiors of these solid tori in $S^3$. The twofold branched cover of each solid torus branched over its core is again a solid torus. To construct the twofold cover of $Z$, we cut along a Seifert surface for the link as we had previously cut along a Seifert surface for the knot, and we complete the construction in an analogous manner.

Those readers with a background in algebraic topology should observe that the twofold branched cover of $S^3$ branched along any simple closed curve $K$ will be a rational homology three-dimensional sphere. In particular, it will always be an orientable three-manifold. However, the twofold branched cover of $S^3$ branched along $K$ will be $S^3$ precisely when $K$ is unknotted.

We can now use twofold branched covers to prove that the graph of a molecular Möbius ladder with three or more rungs is topologically chiral. Recall that the molecular Möbius ladder is embedded in $\mathbb{R}^3$ in the form of a Möbius strip, and we call this embedding the standard embedding of $M_n$.

**Theorem 3.2.** *Let* $M_n$ *be a Möbius ladder that is standardly embedded in* $S^3$, *with* $n \geqslant 3$ *rungs and loop* K. *Then there is no orientation-reversing homeomorphism* $h : (S^3, M_n) \to (S^3, M_n)$ *such that* $h(K) = K$ (*Simon; 1986*).

*Proof.* As a result of the particular embedding of $M_n$ that we are working with, there is an orientation-preserving homeomorphism $g : (S^3, M_n) \to (S^3, M_n)$ that takes each vertex $i$ to the subsequent vertex $i + 1$ on $K$. The reader can see this homeomorphism by rotating $M_n$ around a central axis perpendicular to the page while twisting the ladder in order to keep the crossing in the same place. We illustrate $g$ in Figure 3.21.

88          3. Möbius Ladders and Related Molecular Graphs

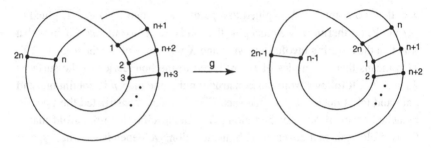

**Figure 3.21.** The homeomorphism $g$ cyclically permutes the vertices of $M_n$

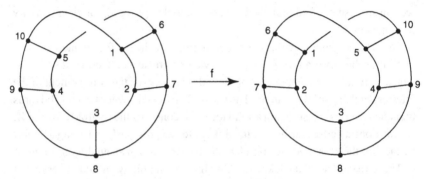

**Figure 3.22.** The homeomorphism $f$ reverses the order of the vertices on $K$

There is also an orientation-preserving homeomorphism $f:(S^3, M_n) \to (S^3, M_n)$ that reverses the orientation of $K$, reversing the order of the vertices around $K$. We obtain $f$ by turning the Möbius ladder over left to right. We illustrate $f$ in Figure 3.22 as a 180° vertical rotation of the graph. If there is an odd number of rungs, one rung will be fixed by this homeomorphism, and if there is an even number of rungs, none will be fixed.

Now we suppose that there does exist an orientation-reversing homeomorphism $h:(S^3, M_n) \to (S^3, M_n)$ such that $h(K) = K$. The consecutive numbering of the vertices around $K$ gives $K$ an orientation. We know that $h$ takes vertices to vertices on $K$ in such a way that adjacent vertices are taken to adjacent vertices. So $h$ either rotates $K$ and cyclically permutes the vertices around $K$, or $h$ reverses the orientation of $K$ and reverses the order of the vertices around $K$. By composing $h$, if necessary, with $f$, we obtain an orientation-reversing homeomorphism of $(S^3, M_n)$ that preserves the orientation of $K$. By further composing $h$ with some power of $g$, we obtain an orientation-reversing homeomorphism of $(S^3, M_n)$ that fixes every vertex. By an abuse of notation

**Figure 3.23.** A standardly embedded $M_3$ can be deformed to this position

we shall still call this homeomorphism $h$. We now consider the Möbius ladder $M_3$ that is obtained from $M_n$ by omitting all but the first three rungs of $M_n$. Because $h$ fixes every vertex of $M_n$, we have $h(M_3) = M_3$. Now $h$ is an orientation-reversing homeomorphism of $(S^3, M_3)$ that fixes all six vertices of $M_3$, and $h(K) = K$.

Because $M_3$ is standardly embedded in $S^3$ in the form of a Möbius strip, the loop $K$ is unknotted. For convenience we deform $K$ to a round planar circle, so that $M_3$ now looks like the graph in Figure 3.23 or its mirror image.

Consider the twofold branched cover of $S^3$ branched along $K$. Because $K$ is unknotted, this branched cover is just $S^3$. Let $p : S^3 \to S^3$ denote the projection map. Let $\overline{M}_3$, $\overline{K}$, $\bar{\alpha}_1$, $\bar{\alpha}_2$, and $\bar{\alpha}_3$ denote the inverse images of $M_3$, $K$, and the rungs $\alpha_1$, $\alpha_2$, and $\alpha_3$, respectively, under the map $p$. That is, $\overline{M}_3 = p^{-1}(M_3) = \{a \in S^3 | p(a) \in M_3\}$, and $\overline{K} = p^{-1}(K) = \{a \in S^3 | p(a) \in K\}$. The covering involution rotates a tube around $\overline{K}$ meridionally, fixing every point of $\overline{K}$. Because the rungs, $\alpha_1$, $\alpha_2$, and $\alpha_3$ have their endpoints on the branch set $K$, the inverse image of each rung is a simple closed curve in the twofold branched cover. With some thought, the reader can see that $\bar{\alpha}_1$, $\bar{\alpha}_2$, and $\bar{\alpha}_3$ are the linked simple closed curves in the twofold branched cover that are illustrated in Figure 3.24 or the mirror image of this picture. In particular, the linking number (as defined in Chapter 2) of any pair of these curves is $\pm 1$, depending on how the curves are oriented.

Because the orientation-reversing homeomorphism $h$ of $(S^3, M_3)$ leaves the branch set $K$ setwise invariant, it is possible to define a homeomorphism $\bar{h}$ of the twofold branched cover that corresponds to $h$, in the sense that $\bar{h} : S^3 \to S^3$ has the property that the composition $p \circ \bar{h}$ is equal to the composition $h \circ p$. Thus given any pair of points $a$ and $b$ in the cover that are both mapped by $p$ to the same point $c$ in the base space, $\bar{h}(a)$ and $\bar{h}(b)$ will both be mapped by $p$ to the point $h(c)$. Now it follows from the properties of $h$ that $\bar{h}(\overline{M}_3) = \overline{M}_3$, $\bar{h}(\overline{K}) = \overline{K}$, $\bar{h}$ fixes every vertex of $\overline{M}_3$, and $\bar{h}$ is orientation reversing. Thus $\bar{h}(\bar{\alpha}_i) = \bar{\alpha}_i$ for each $i$.

We choose orientations for $\bar{\alpha}_1$, $\bar{\alpha}_2$, and $\bar{\alpha}_3$ such that $\text{Lk}(\bar{\alpha}_1, \bar{\alpha}_2) = 1$ and $\text{Lk}(\bar{\alpha}_1, \bar{\alpha}_3) = 1$. From the picture it can be seen that, with these orientations,

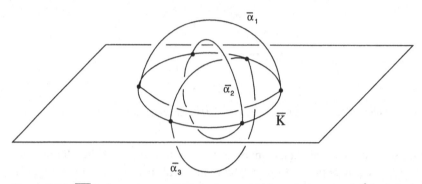

**Figure 3.24.** $\overline{M_3}$ is the preimage of $M_3$ in the twofold branched cover of $S^3$ branched over the loop $K$

$Lk(\bar{\alpha}_2, \bar{\alpha}_3) = 1$. Recall from Chapter 2 that because $\bar{h} : S^3 \rightarrow S^3$ is an orientation-reversing homeomorphism, we have $Lk[\bar{h}(\bar{\alpha}_i), \bar{h}(\bar{\alpha}_j)] = -Lk(\bar{\alpha}_i, \bar{\alpha}_j)$. First we assume that $\bar{h}$ preserves the orientation of $\bar{\alpha}_1$; then $Lk[\bar{h}(\bar{\alpha}_1), \bar{h}(\bar{\alpha}_2)] = Lk[\bar{\alpha}_1, \bar{h}(\bar{\alpha}_2)] = -1$, and $Lk[\bar{h}(\bar{\alpha}_1), \bar{h}(\bar{\alpha}_3)] = Lk[\bar{\alpha}_1, \bar{h}(\bar{\alpha}_3)] = -1$. So, $\bar{h}$ must reverse the orientation of both $\bar{\alpha}_2$ and $\bar{\alpha}_3$, so $Lk[\bar{h}(\bar{\alpha}_2), \bar{h}(\bar{\alpha}_3)] = Lk(-\bar{\alpha}_2, -\bar{\alpha}_3)$. But we know from the properties of the linking number explained in Chapter 2 that $Lk(-\bar{\alpha}_2, -\bar{\alpha}_3) = Lk(\bar{\alpha}_2, \bar{\alpha}_3)$. So $Lk(\bar{h}(\bar{\alpha}_2), \bar{h}(\bar{\alpha}_3)) = -1$. This is a contradiction since $Lk(\bar{h}(\bar{\alpha}_2), \bar{h}(\bar{\alpha}_3)) = -Lk(\bar{\alpha}_2, \bar{\alpha}_3) = 1$. We get a similar contradiction by assuming that $\bar{h}$ reverses the orientation of $\bar{\alpha}_1$. Hence no such homeomorphism $h$ can exist. $\qquad\square$

Theorem 3.2 tells us that the graph of any molecular Möbius ladder with three or more rungs must be topologically chiral in $S^3$ and hence in $\mathbb{R}^3$. In particular, the three-rung molecular Möbius ladder synthesized by Walba's lab in 1982 (Walba et al., 1982) and the four-rung Möbius ladder synthesized by Walba's lab in 1986 (Walba et al., 1986) must both be topologically chiral and therefore chemically chiral.

### Intrinsic Chirality

The proof of Simon's Theorem uses both the abstract structure of the graph of $M_n$ and the particular embedding of $M_n$ that looks like a Möbius strip. We saw in Chapter 1 that if $n$ is even, then there is a different embedding of $M_n$ in $\mathbb{R}^3$ that is topologically achiral. Figure 3.25 reproduces the illustration from Chapter 1 of a topologically achiral embedding of $M_4$. If this graph is rotated about a

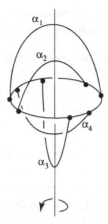

**Figure 3.25.** A topologically achiral embedding of $M_4$

vertical axis by 90°, we will obtain its mirror image. We can add symmetric pairs of edges at the top and the bottom of the graph in such a way that, for any even number $n$, we have a topologically achiral embedding of $M_n$ in $\mathbb{R}^3$.

We will prove below that for $n \geqslant 3$ odd, if we distinguish between the edges that are rungs and the edges that are sides, then every embedding of $M_n$ is topologically chiral. First, we give a different definition of the linking number that will be more useful to us here. There are many equivalent definitions of the linking number that are each useful in different settings (see, e.g., Rolfsen, 1976). We state our definition in $S^3$; however, it could equivalently be stated in $\mathbb{R}^3$.

**Definition.** Let $J$ and $K$ be disjoint oriented simple closed curves in $S^3$. Let $S$ denote a two-sided surface bounded by $K$. Label the side of $S$ where $K$ is oriented counterclockwise as the positive side, and the side of $S$ where $K$ is oriented clockwise as the negative side. We follow along the oriented curve $J$, and for each point where $J$ passes through $S$ going from the negative to the positive side of $S$ we add $+1$, and for each point where $J$ passes through $S$ going from the positive to the negative side of $S$ we add $-1$. This sum over all points where $J$ passes through $S$ is defined to be $\mathrm{Lk}(K, J)$.

We have seen that every simple closed curve in $S^3$ bounds a two-sided surface. It can be shown that the linking number defined as above is independent of the choice of the particular surface $S$. Furthermore, this definition can be proven to be equivalent to the definition of the linking number that we gave in Chapter 2 and hence is a topological invariant of an oriented link with components $J$ and

$K$ (Rolfsen, 1976). In particular, it follows that Lk$(J, K) =$ Lk$(K, J)$. Recall from Chapter 2 that the linking number has the following additional properties.

1. For any orientation-reversing homeomorphism $h : S^3 \to S^3$, we have
   Lk$(J, K) = -$Lk$[h(J), h(K)]$.
2. Lk$(J, K) = -$Lk$(-J, K) = -$Lk$(J, -K) =$ Lk$(-J, -K)$, where
   a minus sign in front of a component means that the orientation of that component has been reversed.

It can also be shown that this definition of the linking number is well defined in the twofold branched cover $M$ of $S^3$ branched over any knot and that it still has all of the above properties. This follows from homology theory and the fact that $H_1(M)$ is finite. See Rolfsen (1976) for a proof that $H_1(M)$ is always finite. We shall use these properties of the linking number in order to prove Theorem 3.3. This theorem will then enable us to prove that any embedding of a Möbius ladder with an odd number of rungs is topologically chiral if we distinguish between the edges that are rungs and the edges that are sides of the ladder. Note that in contrast with Theorem 3.2, the statement of Theorem 3.3 (Flapan, 1989) does not assume that $M_n$ is embedded in the form of a Möbius strip. In this case, there are no restrictions at all on the embedding of $M_n$. So, for example, the loop $K$, together with any of the rungs, may contain knots. It follows from Lemma 3.1 that no matter how $M_n$ is embedded, if $n > 3$, then any homeomorphism $h : (S^3, M_n) \to (S^3, M_n)$ will necessarily have the property that $h(K) = K$. Thus, as in Theorem 3.2, our hypothesis that $h(K) = K$ is actually only required when $n = 3$.

**Theorem 3.3.** *Let* n $\geqslant$ 3 *and let* M$_n$ *be a Möbius ladder that is embedded in* S$^3$. *Suppose that* h : (S$^3$, M$_n$) $\to$ (S$^3$, M$_n$) *is an orientation-reversing homeomorphism such that* h(K) = K, *where* K *is the loop of* M$_n$. *Then there is no rung* $\alpha_i$ *such that* h($\alpha_i$) = $\alpha_i$.

*Proof.* Let $X$ denote the twofold branched cover of $S^3$ branched over $K$. Because we do not know whether or not $K$ is knotted, we do not know whether or not $X$ is actually $S^3$. However, in any case, the linking number is well defined in $X$ and has all of the usual properties. Let $\overline{K}$ denote the inverse image of $K$ under the projection map $p : X \to S^3$.

Observe that for any rung $\alpha_i$, the cut open loop, $K - \alpha_i$, always has two components. For each $i$, we shall let $k_i$ denote the simple closed curve in $S^3$ consisting of the rung $\alpha_i$ together with one component of $K - \alpha_i$ (it does not matter which component). Let $F_i$ be a Seifert surface for $k_i$. For each $i$, let $S_i$ denote the inverse image of $F_i$ and let $K_i$ denote the inverse image of $\alpha_i$ under $p$.

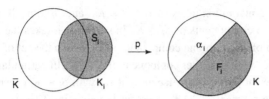

**Figure 3.26.** $S_i$ is a surface bisected by $\overline{K}$ such that each component of $S_i - \overline{K}$ gets mapped to $F_i$

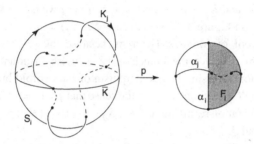

**Figure 3.27.** $K_j$ intersects $S_i - \overline{K}$ in an even number of points

Thus $p(S_i) = F_i$ and $p(K_i) = \alpha_i$. The boundary of the surface $S_i$ is the simple closed curve $K_i$. Note that because $K$ is the branch set and $F_i$ contains an arc of $K$ in its boundary, $\overline{K}$ cuts each $S_i$ in half. Figure 3.26 illustrates $S_i$ as a surface bisected by $\overline{K}$ such that each component of $S_i - \overline{K}$ gets mapped to $F_i$.

Because each $F_i$ is an orientable surface, each $S_i$ is also an orientable surface. For each $i$, we give $K_i$ an orientation. Now $S_i$ is two sided because it is an orientable surface embedded in the orientable three-manifold $X$. Label the positive and negative side of $S_i$ as indicated in the definition of the linking number. Thus for each distinct pair of simple closed curves $K_i$ and $K_j$ in $X$, the linking number $\mathrm{Lk}(K_i, K_j)$ is equal to the sum obtained by adding $+1$ for each point where $K_j$ passes through $S_i$ going from the negative to the positive side, and $-1$ for each point where $K_j$ passes through $S_i$ going from the positive to the negative side.

Figure 3.27 is a schematic representation of $K_j$ and $S_i$ in the twofold branched cover. Here $S_i$ is the vertical hemisphere that faces us, and $\overline{K}$ is a horizontal circle that bisects the hemisphere $S_i$. The covering involution rotates the space around $\overline{K}$, exchanging the upper half-hemisphere of $S_i$ with the lower half-hemisphere of $S_i$, and exchanging the interior of the ball with the exterior of the ball. In the example of Figure 3.27, the simple closed curve $K_j$ intersects $\overline{K}$ in two points, only one of which is on $S_i$, and $K_j$ intersects each component of $S_i - \overline{K}$ in two points.

The covering involution takes each of $K_i$, $K_j$, and $S_i$ to itself. It also inter-changes the two components of $S_i - \overline{K}$. In so doing, it takes the set of points of intersection of $K_j$ with one component of $S_i - \overline{K}$ to the set of points of intersection of $K_j$ with the other component of $S_i - \overline{K}$. In particular, this means that the number of points of intersection of $K_j$ with $S_i - \overline{K}$ is even. For any $i$ and $j$, the rung $\alpha_j$ meets $F_i \cap K$ in precisely one point. Because the covering involution fixes every point of $K$, the projection map $p$ restricted to $\overline{K}$ is one to one. Hence, there is precisely one point where $K_j$ meets $S_i \cap \overline{K}$. Thus, the total number of points of intersection of $K_j$ with $S_i$ is odd. In particular, no matter how we orient $K_i$ and $K_j$, this means that for all $i \neq j$ we have $\mathrm{Lk}(K_i, K_j) \neq 0$. For our example in Figure 3.27, we have $\mathrm{Lk}(K_i, K_j) = -1$.

As in the proof of Simon's Theorem, because the orientation-reversing homeomorphism $h$ leaves the branch set $K$ setwise invariant, we can define an orientation-reversing homeomorphism of the twofold branched cover, $g : X \to X$, which corresponds to $h$ in the sense that $p \circ g = h \circ p$. In addition, because $g$ is orientation reversing, $\mathrm{Lk}(K_i, K_j) = -\mathrm{Lk}[g(K_i), g(K_j)]$ for every distinct pair $i$ and $j$.

Now by hypothesis we have $h(K) = K$, so $h$ either rotates the vertices around $K$ or reflects the vertices around $K$. In order to reach a contradiction, we now suppose there exists some $\alpha_i$ such that $h(\alpha_i) = \alpha_i$. Then either $h$ takes each rung to itself, or else for this specific $i$, $h(\alpha_{i-1}) = \alpha_{i+1}$ and $h(\alpha_{i+1}) = \alpha_{i-1}$, where we assume the rungs are labeled consecutively. We consider these two cases separately.

*Case 1.* $h(\alpha_i) = \alpha_i$ for every rung $\alpha_i$.

It follows from the hypothesis of this case that $g(K_i) = K_i$ for every $i$, but we do not know whether $g$ reverses or preserves the orientation of any particular $K_i$. Suppose that $g$ preserves the orientation of $K_1$. Then because $\mathrm{Lk}(K_1, K_2) = -\mathrm{Lk}[g(K_1), g(K_2)] = -\mathrm{Lk}[K_1, g(K_2)]$ and $\mathrm{Lk}(K_1, K_2) \neq 0$, it follows that $g$ reverses the orientation of $K_2$. Similarly, because $\mathrm{Lk}(K_1, K_3) \neq 0$, it follows that $g$ reverses the orientation of $K_3$. However, because $\mathrm{Lk}(K_2, K_3) \neq 0$, $g(K_i) = K_i$ for each $i$, and $\mathrm{Lk}(K_2, K_3) = -\mathrm{Lk}[g(K_2), g(K_3)]$, precisely one of $K_2$ and $K_3$ has its orientation reversed by $g$. This is a contradiction. We obtain a similar contradiction if we start with the assumption that $g$ reverses the orientation of $K_1$. Thus this case cannot arise.

*Case 2.* There exists some $i$ such that $h(\alpha_i) = \alpha_i$, $h(\alpha_{i-1}) = \alpha_{i+1}$, and $h(\alpha_{i+1}) = \alpha_{i-1}$.

In this case, we have $g(K_i) = K_i$, $g(K_{i-1}) = K_{i+1}$, and $g(K_{i+1}) = K_{i-1}$. Because $\mathrm{Lk}(K_{i-1}, K_{i+1}) = -\mathrm{Lk}[g(K_{i-1}), g(K_{i+1})]$, the oriented link of

$K_{i-1}$ and $K_{i+1}$ is the same as the oriented link of $g(K_{i-1})$ and $g(K_{i+1})$ except that the orientation of one component is reversed. Without loss of generality we can assume that $g(K_{i-1}) = +K_{i+1}$, and $g(K_{i+1}) = -K_{i-1}$. This means that, $g(K_{i-1})$ has the same orientation as $K_{i+1}$ and $g(K_{i+1})$ has the opposite orientation as $K_{i-1}$. Suppose that $g$ preserves the orientation of $K_i$. Then $\text{Lk}(K_i, K_{i+1}) = -\text{Lk}[g(K_i), g(K_{i+1})] = -\text{Lk}(K_i, -K_{i-1})$ and $\text{Lk}(K_i, K_{i-1}) = -\text{Lk}[g(K_i), g(K_{i-1})] = -\text{Lk}(K_i, K_{i+1})$. From these equations we can deduce that $\text{Lk}(K_i, -K_{i-1}) = \text{Lk}(K_i, K_{i-1})$. This is impossible since $\text{Lk}(K_i, -K_{i-1}) = -\text{Lk}(K_i, K_{i-1})$ and $\text{Lk}(K_i, K_{i-1}) \neq 0$. We obtain a similar contradiction if $g$ reverses the orientation of $K_i$. Hence this case also does not arise.

Therefore, there is no rung $\alpha_i$ such that $h(\alpha_i) = \alpha_i$. □

We can use Theorem 3.3 to prove the topological chirality of any embedding of a Möbius ladder with an odd number of rungs, no matter how knotted or tangled it may be.

**Corollary 3.4.** *Let* $M_n$ *be a Möbius ladder that is embedded in* $S^3$, *and let* $n \geqslant 3$ *be odd. Then there is no orientation-reversing homeomorphism* $h : (S^3, M_n) \rightarrow (S^3, M_n)$ *with* $h(K) = K$, *where* $K$ *is the loop of* $M_n$ *(Flapan, 1989).*

*Proof.* Suppose that there is such a homeomorphism $h$. By hypothesis $h(K) = K$, so $h$ either rotates or reflects the vertices around $K$. Suppose that $h$ rotates the vertices around $K$. Then $h$ also rotates the rungs about $K$, and there is some minimal number $r$ such that $h^r$ takes each rung to itself. Observe that, because $h$ rotates the rungs around $K$, this number $r$ must divide the number of rungs $n$, so $h^n$ takes each rung to itself. However, as $n$ is odd, $h^n$ is an orientation-reversing homeomorphism of $(S^3, M_n)$ such that $h^n(K) = K$ and $h^n$ takes each rung to itself. Therefore the existence of the homeomorphism $h^n$ contradicts Theorem 3.3.

Hence $h$ cannot rotate the vertices around $K$, so it must reflect the vertices around $K$. This means that $h$ induces an order-two automorphism on the vertices of $K$, and hence on the rungs. However, because there is an odd number of rungs, there must be at least one rung $\alpha_i$ such that $h(\alpha_i) = \alpha_i$. Again, the existence of the orientation-reversing homeomorphism $h$ contradicts Theorem 3.3. Therefore, there cannot exist an orientation-reversing homeomorphism $h(S^3, M_n) \rightarrow (S^3, M_n)$ with the property that $h(K) = K$. □

This corollary motivates the following definition.

**Figure 3.28.** $[m][n][p]$-paracyclophane is intrinsically chiral

**Definition.** A labeled or unlabeled graph $G$ is said to be *intrinsically chiral* if every embedding of $G$ in $\mathbb{R}^3$ is topologically chiral.

Recall that a graph is topologically chiral in $S^3$ if and only if it is topologically chiral in $\mathbb{R}^3$. Thus a graph is intrinsically chiral in $S^3$ if and only if it is intrinsically chiral in $\mathbb{R}^3$.

If we consider the edges of $M_n$ to be labeled or colored so that the sides are distinguished from the rungs, then Corollary 3.4 says that if $n \geqslant 3$ is odd then $M_n$ is intrinsically chiral. If we do not label the edges, then by Lemma 3.1 and Corollary 3.4, $M_n$ is intrinsically chiral if $n > 3$ is odd. From a chemical point of view, when the sides of a Möbius ladder are necessarily distinct from the rungs, Corollary 3.4 implies that every real or hypothetical topological stereoisomer of a molecular Möbius ladder with an odd number of rungs is chemically chiral.

There can be other molecules whose structure has the form of a Möbius ladder with three rungs. For example, Figure 3.28 illustrates the graph of a $[m][n][p]$-paracyclophane, proposed by Nakazaki (1984). Here $m$, $n$, and $p$ indicate the number of copies of a particular molecular subunit (for example, $CH_2$) that occur on the molecular chain. The graph in Figure 3.28 is an $M_3$ where the sides of the ladder represent the benzene ring and the rungs represent the $m$, $n$, and $p$ chains. In particular, this means that the sides are distinguished from the rungs. Hence $[m][n][p]$-paracyclophane is also intrinsically chiral. In fact, this particular embedding of the graph of $[m][n][p]$-paracyclophane can be deformed to a standardly embedded Möbius ladder, in the form of a Möbius strip.

Liang and Mislow (1994b) have observed that some proteins also have the form of an $M_3$. For example, Figure 3.29 illustrates the protein variant-3 scorpion neurotoxin from *Centruroides sculpturatus Ewing*. In order to explain this picture we begin with a little background about protein molecules. Proteins are made up of repeating patterns of amino acids that are joined together by *peptide bonds*. A chain of amino acids joined together by peptide bonds is called a *polypeptide chain*. One end of a polypeptide chain always terminates with $NH_3^+$ and the other end of the chain always terminates with $CO_2^-$. In order to distinguish these two ends, the former is referred to as the *N terminal*, and the latter is referred to as the *C terminal*. In Figure 3.29, the polypeptide chain

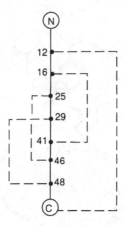

**Figure 3.29.** This protein is intrinsically chiral

is the vertical line segment from the N to the C terminals. The dashed lines indicate *intrachain disulfide bonds*, that is, bonds joining one amino acid on the polypeptide chain to another amino acid on the chain. The numbers on the line segment enable us to see at what point along the polypeptide chain the amino acids are bonded together. For example, we see from the picture that the sixteenth and the forty-first amino acids are joined together by disulfide bonds.

We can see the Möbius ladder in this graph as follows. Let $K$ denote the simple closed curve going from the 12 along the dashed disulfide bond to the terminal C, then along the polypetide chain through all of the numbered vertices back to the 12. The remaining three disulfide bonds represent the rungs of the Möbius ladder. This protein is actually an $M_3$ with one extra vertex (as represented by the terminal $N$) and one extra edge attaching the $N$ to the rest of the graph. The rungs can be chemically distinguished from the loop $K$, because the rungs are precisely those three disulfide bonds that do not contain the terminal C. Any homeomorphism of this graph to itself will take this $M_3$ to itself setwise, and take the terminal N to itself. It follows from Corollary 3.4 that the protein illustrated in Figure 3.29 is intrinsically chiral.

Observe that not every molecular graph that has the general form of a Möbius ladder is necessarily intrinsically chiral. Figure 3.30 illustrates the graph of the molecule Kuratowski cyclophane, which has the general form of an $M_3$. However, only the central rung in Kuratowski cyclophane is distinct from the sides of the ladder. In particular, the two diagonal rungs are chemically indistinguishable from the sides of the ladder. This graph can actually be deformed so that it looks like the topologically achiral $M_3$ that was illustrated in Figure 3.2 of this chapter; so this molecular graph is topologically achiral as well. Chen et al.

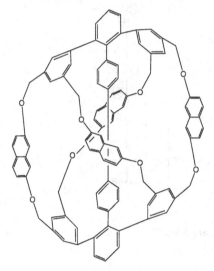

**Figure 3.30.** Kuratowski cyclophane is topologically achiral

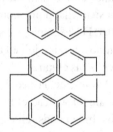

**Figure 3.31.** Triple-layered naphthalenophane

(1995) designed and synthesized Kuratowski cyclophane, and they showed that on a chemical level it is also achiral.

## Triple-Layered Naphthalenophane

There exist various other types of molecules whose molecular bond graphs are related to the graph of a Möbius ladder. For example, let us consider the graph of the triple-layered naphthalenophane molecule, which is illustrated in Figure 3.31 (this molecule was synthesized by Otsubo, Ogura, & Misumi, (1983)). As usual, the hexagons in the illustration represent benzene rings. A pair of benzene rings that are fused together (attached along a side of the hexagon) is called a *naphthalene*. Roughly speaking, this graph has a structure

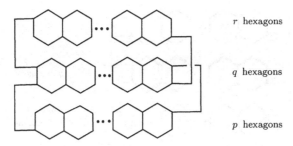

r hexagons

q hexagons

p hexagons

**Figure 3.32.** The graph $G(p, q, r)$

that is similar to $M_3$, where the ten edges around the central naphthalene play the role of the loop $K$. The edge shared by the two central hexagons plays the role of one rung of the $M_3$, and the top naphthalene, which is joined by edges to $K$, and the bottom naphthalene, which is joined by edges to $K$, play the roles of the two additional rungs of $M_3$. Liang and Mislow (1994a) observed that if the bottom pair of hexagons is replaced by a single edge and the top pair of hexagons is also replaced by a single edge, then we will actually have an $M_3$ (together with some extra vertices), and they concluded that the graph of triple-layered naphthalenophane is intrinsically chiral. However, this does not provide a proof of the intrinsic chirality of the graph, because the definition of topological achirality does not permit us to replace a simple closed curve with a single edge. Furthermore, the proof of Theorem 3.3 does not readily adapt to this graph. Nonetheless, we can prove that triple-layered naphthalenophane is intrinsically chiral by making use of Corollary 3.4. In fact, we can actually prove a more general result than this.

We create a class of graphs that are similar to the graph of triple-layered naphthalenophane by fusing together any number of hexagons in place of each of the naphthalenes in the graph of triple-layered naphthalenophane. Figure 3.32 illustrates an embedding of such a graph, with $p$ hexagons fused together on the bottom, $q$ hexagons fused together in the middle, and $r$ hexagons fused together on the top. We will denote the abstract graph that is embedded in Figure 3.32 by $G(p, q, r)$. Note that each corner in the graph represents a vertex. Observe that the graph of triple-layered naphthalenophane is $G(2, 2, 2)$. We will prove below that if $p, q$, and $r$ are all even, then $G(p, q, r)$ is intrinsically chiral (Flapan & Forcum, 1998). Thus, in particular, the graph of triple-layered naphthalenophane is intrinsically chiral.

**Theorem 3.5.** *If* p, q, *and* r *are even, then the graph* G(p,q,r) *is intrinsically chiral.*

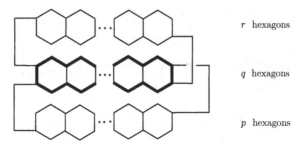

r hexagons

q hexagons

p hexagons

**Figure 3.33.** The circuit $K$ is darkened in $G(p, q, r)$

*Proof.* We shall prove this by contradiction. Suppose that $p$, $q$, and $r$ are all even, and there is a topologically achiral embedding of $G(p, q, r)$ in $S^3$. We shall show that this assumption implies that there is an embedding of $M_3$ in $S^3$ such that there is an orientation-reversing homeomorphism $h : S^3 \rightarrow S^3$ with $h(M_3) = M_3$ and $h(K) = K$. By our assumption there is an orientation reversing homeomorphism $h : S^3 \rightarrow S^3$ with $h[G(p, q, r)] = G(p, q, r)$. The embedding of $G(p, q, r)$ may not look at all like that of Figure 3.32; however, as we do not know how $G(p, q, r)$ is embedded, it is convenient to draw it as in Figure 3.32. However, note that our proof does not rely on this particular embedding. Now $h$ takes any simple closed curve with six vertices to a simple closed curve with six vertices. Because the only simple closed curves with six vertices are the obvious hexagons, this means that $h$ takes each of these hexagons to itself or to another such hexagon. Furthermore, $h$ must take any pair of adjacent hexagons to a pair of adjacent hexagons. So $h$ must take each chain of fused hexagons to a chain of the same number of fused hexagons. The chain of $q$ hexagons is the only chain of hexagons where the outer two hexagons each contain four vertices of valence three. So, $h$ must take this chain of $q$ hexagons to itself, possibly switching the two sides or the top and the bottom. Let $K$ denote the perimeter of the chain of $q$ hexagons. Then $K$ contains $4q + 2$ edges, because each hexagon in the chain contributes four edges to the perimeter, except the terminal hexagons that each contribute five edges to $K$. It follows that $h$ must take the set $K$ to itself, though $h$ may interchange the top and the bottom or the left and the right sides of $K$. In Figure 3.33 we illustrate $G(p, q, r)$ with $K$ as a darkened simple closed curve.

It follows from the above paragraph that $h$ must either take the chain of $p$ hexagons to itself and take the chain of $r$ hexagons to itself, or if $r = p$, then $h$ might interchange these two chains of hexagons. Let $x$, $y$, $z$, and $w$ denote the vertices indicated in Figure 3.34. Because $h$ takes $K$ to itself and takes the other chains of hexagons to themselves or to each other, $h$ must take the set of points

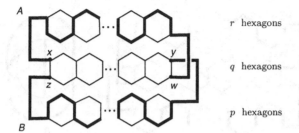

A

*r* hexagons

*q* hexagons

*p* hexagons

B

**Figure 3.34.** The arcs *A* and *B* are darkened in $G(p, q, r)$

$\{x, y, w, z\}$ to itself, either leaving $\{x, w\}$ and $\{y, z\}$ setwise invariant or else interchanging the pair $\{x, w\}$ with the pair $\{y, z\}$. Recall that by hypothesis, *r* is even, so the *r* hexagons in that chain occur in pairs. Thus the arc that zigzags from left to right through the chain of *r* hexagons going down first and then up (as in Figure 3.34) will contain two more edges than the arc that zigzags from left to right through the *r* hexagons in the opposite way (initially going up rather than down). In particular, as *r* is even, there is a unique non-self-intersecting arc *A* in $G(p, q, r) - K$ with endpoints at *x* and *w* such that *A* contains $7 + 3r$ edges. Similarly, as *p* is even, there is a unique non-self-intersecting arc *B* in $G(p, q, r) - K$ with endpoints at *y* and *z* such that *B* contains $7 + 3p$ edges. The arcs *A* and *B* are darkened in Figure 3.34. Now *h* either sends each of the arcs *A* and *B* to itself or possibly interchanges them when $r = p$.

Because *q* is even, there exists a unique edge *C* that splits the chain of *q* hexagons in half, that is, such that each component of $K - C$ contains half of the vertices of $K - C$. Because *C* is the unique edge with this property and $h(K) = K$, it follows that $h(C) = C$.

Let *M* denote the graph obtained by taking the union of the circuit *K* together with the arcs *A*, *B*, and *C*. We saw that $h(K) = K$, $h(C) = C$, and *h* either interchanges *A* and *B* or takes each of *A* and *B* to itself. In either case, it follows that $h(M) = M$. We illustrate *M* on the left side of Figure 3.35. On the right side of Figure 3.35 we have drawn *M* in a more normal way as an embedded three-rung Möbius ladder with loop *K* and rungs *A*, *B*, and *C*.

Thus $h : (S^3, M) \to (S^3, M)$ is an orientation-reversing homeomorphism with $h(K) = K$. This contradicts Corollary 3.4. It follows that if *p*, *q*, and *r* are all even, then the graph $G(p, q, r)$ is intrinsically chiral. □

Now we want to consider the graph $G(p, q, r)$ where at least one of *p*, *q*, or *r* is odd. We shall show that in this case, $G(p, q, r)$ has an embedding that is topologically achiral, and hence the graph $G(p, q, r)$ is not intrinsically chiral.

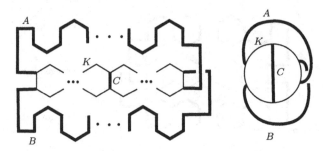

**Figure 3.35.** *M* is a three–rung Möbius ladder

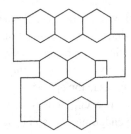

**Figure 3.36.** The standard embedding of the graph $G(3, 2, 2)$

First we consider the case in which at least one of $p$ or $r$ is odd and $q$ is either even or odd. We shall illustrate a topologically achiral embedding of $G(3, 2, 2)$, and then see how to generalize it to obtain a topologically achiral embedding of any $G(p, q, r)$ where $p$ or $r$ is odd. Figure 3.36 illustrates a standard embedding of the graph $G(3, 2, 2)$.

In Figure 3.37 we illustrate a topologically achiral embedding of $G(3, 2, 2)$. From a chemical point of view, of course, one benzene ring cannot pass through another, but if we consider $G(3, 2, 2)$ as an abstract graph, there is nothing wrong with the embedding illustrated in Figure 3.37. In this figure, the chain of three hexagons is contained in a vertical plane, while the rest of the graph lies in a horizontal plane. The round arc at the top is actually the shared edge of the two largest hexagons. To obtain the mirror image we first rotate the figure by 180° about a vertical axis going through the center of the three vertical hexagons and the shared edge of the two fused small hexagons. This rotation takes the horizontal plane to itself and takes the graph to its mirror image except for the circular arc, which is now at the bottom instead of the top of the picture. To correct this, we swing the circular arc over the rest of the figure so that it goes back to its original position. In this way we can deform the graph to its mirror image.

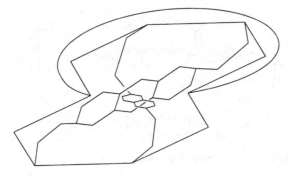

**Figure 3.37.** A topologically achiral embedding of $G(3, 2, 2)$

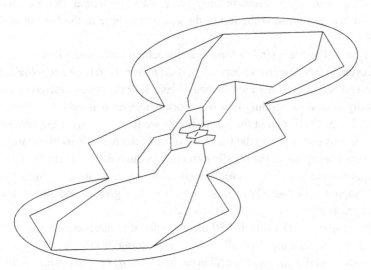

**Figure 3.38.** A topologically achiral embedding of $G(3, 3, 3)$

We have a similar topologically achiral embedding for any $G(p, q, r)$ where $p$ or $r$ is odd. We can see from Figure 3.37 that increasing the number of horizontal hexagons running through the center of the figure has no effect on the symmetry, as long as the number of vertical hexagons is odd. If $q$, the number of large hexagons, is odd, we will end up with an even number of circular arcs; thus we will obtain the mirror image just by rotating the figure by 180° without having to swing a circular arc from the bottom to the top. We illustrate this with a topologically achiral embedding of $G(3, 3, 3)$ in Figure 3.38. This embedded graph can be rotated by 180° to obtain its mirror image. For any $q$ that is even, we will have a circular arc, whose endpoints occur at the center of the row of large hexagons, that will have to be swung from the bottom

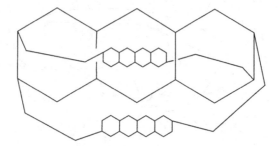

**Figure 3.39.** A topologically achiral embedding of $G(4, 3, 4)$

to the top of the graph after rotating. The remaining circular arcs are evenly divided between those at the top of the graph and those at the bottom of the graph.

We cannot create a similar topologically achiral embedding if $p$ and $r$ are both even. In this case, by Theorem 3.5, if $G(p, q, r)$ is to have a topologically achiral embedding, then $q$ must be odd. Figure 3.39 illustrates a different topologically achiral embedding that will work whenever $q$ is odd. We illustrate it for $G(4, 3, 4)$. To obtain the mirror image we first rotate the figure by 180° about an axis that is perpendicular to the plane of the paper. This will cause the four small hexagons in the middle to remain in the middle and the four small hexagons at the bottom to go to the top. Then swing the arc containing these top hexagons back to the bottom of the picture. This gives us the mirror image of the graph through the plane of the paper.

Thus, Figures 3.37 through 3.39 illustrate that if at least one of $p$, $q$, or $r$ is odd, then there is a topologically achiral embedding of $G(p, q, r)$ in $\mathbb{R}^3$. We saw above that if $p$, $q$, and $r$ are all even then $G(p, q, r)$ is intrinsically chiral. Thus we have completely characterized when $G(p, q, r)$ is intrinsically chiral. The reader should observe that, regardless of the values of $p$, $q$, and $r$, the graph $G(p, q, r)$ contains a Möbius ladder with three rungs, where the perimeter of the $q$ central hexagons represents the loop $K$ of the Möbius ladder. However, in the cases in which there is a topologically achiral embedding of $G(p, q, r)$, no $M_3$ that is contained in $G(p, q, r)$ will be invariant under an orientation-reversing homeomorphism of $[\mathbb{R}^3, G(p, q, r)]$. Rather, any $M_3$ will get mapped to another $M_3$ that is also contained in the graph. We can see from these examples that just because a graph contains an $M_3$ in which the rungs can be distinguished from the edges does not mean that the graph itself is intrinsically chiral. Such an $M_3$ will enable us to show that a graph $G$ is intrinsically chiral only if we can show that this Möbius ladder must be invariant under any orientation reversing homeomorphism of $(\mathbb{R}^3, G)$.

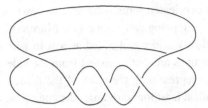

**Figure 3.40.** The sides of a closed ladder with three half-twists form a trefoil knot

It should be observed that all of the examples illustrated in Figures 3.37 through 3.39 have the property that a piece of the graph passes through one of the hexagons. It will be shown in Chapter 4 that if each of the hexagons of $G(p, q, r)$ is filled in with a disk so that no piece of the graph can pass through it, then the structure we obtain will be intrinsically chiral.

## Multiply Twisted Closed Ladders

The molecular Möbius ladders are closed ladders with one half-twist. Walba used similar chemical techniques to attempt to synthesize closed ladders with multiple half-twists, with the intention of then removing the carbon–carbon double bonds to obtain a knot. In Figure 3.40 we illustrate the trefoil knot as a Möbius ladder with three half-twists where the rungs have been removed. Although Walba did not succeed in synthesizing multiply twisted closed ladders, the attempt led him to conjecture that every closed ladder with at least three rungs, embedded in the form of a strip with any positive number of half-twists, is topologically chiral if we distinguish between the edges that are sides and the edges that are rungs. By Simon's Theorem, this was known for ladders with one half-twist. A closed ladder with no half-twists has the form of a cylinder and is topologically achiral. Any closed ladder with an odd number of half-twists is an embedding of a Möbius ladder. So, if we distinguish between the edges that are sides and the edges that are rungs, then any closed ladder with an odd number of twists and an odd number of rungs will be topologically chiral by Corollary 3.4.

Walba's conjecture also turns out to be correct for any closed ladder with at least three half-twists regardless of the number of rungs. To see this, first observe that the sides of such a ladder form a torus knot or link. Recall from Chapter 2 that a *torus knot* or *torus link* is defined to be any nontrivial knot or link that can be embedded in the surface of a torus. For example, in Figure 3.40 we can imagine inserting a torus so that the trefoil knot lies on the surface of the torus, twisting around it three times meridionally and two times longitudinally.

All torus knots and unoriented torus links with at least three crossings have long been known to be topologically chiral (see Murasugi, 1987, for a proof). Let $G$ represent a closed ladder embedded as a strip with at least three half-twists, in which the rungs are distinguished from the sides. Let $L$ denote the simple closed curves representing the sides of the ladder. Then $L$ is either a torus knot or link with at least four crossings, depending on whether there is an odd or an even number of half twists in the ladder. If there were an orientation-reversing homeomorphism $h$ of $(S^3, G)$ such that $h(L) = L$, it would contradict the topological chirality of the torus knots and links. Thus any closed ladder embedded as a strip with at least three half-twists must be topologically chiral.

We cannot use this argument if the ladder has only two half-twists because then the sides of the ladder form a Hopf link that can be deformed to its mirror image. Simon (1987) proved, as follows, that the closed ladders with two half-twists are topologically chiral if they have at least three rungs. Let $n > 2$ and let $C_n$ be a closed ladder with $n$ rungs embedded in the form of a cylinder with two half-twists, in which rungs are distinguished from sides. Observe that the sides of $C_n$ consist of two simple closed curves, which we shall denote by $A$ and $B$. Let the vertices on $A$ be denoted by $a_1, \ldots, a_n$ and let the vertices on $B$ be denoted by $b_1, \ldots, b_n$ in such a way that each rung $\alpha_i$ has vertices $a_i$ and $b_i$. Suppose there is an orientation-reversing homeomorphism $h : (S^3, C_n) \to (S^3, C_n)$ with $h(A \cup B) = A \cup B$. That is, $h$ takes the sides to the sides and the rungs to the rungs. We make no assumption about whether or not $h$ interchanges the simple closed curves $A$ and $B$. Because of the particular embedding of $C_n$, there is an orientation-preserving homeomorphism $g$ that rotates $C_n$ while twisting it, taking each $a_i$ to $a_{i+1}$ and each $b_i$ to $b_{i+1}$. There is also an orientation-preserving homeomorphism $f$ that rotates $C_n$ about a horizontal axis, interchanging the simple closed curves $A$ and $B$. The homeomorphisms $g$ and $f$ are illustrated in Figure 3.41 for $C_3$. Note that both $f$ and $g$ are orientation preserving.

By composing $h$, if necessary, with $f$ or some power of $g$, we can obtain an orientation-reversing homeomorphism that we call $j$, such that $j(A) = A$ and $j(B) = B$, and $j(a_1) = a_1$ and $j(b_1) = b_1$. Suppose that $C_n$ has at least three rungs. Then either $j(a_i) = a_i$ and $j(b_i) = b_i$ for each $i$, or $j(a_2) = a_n$ and $j(b_2) = b_n$. We can orient $A$ and $B$ by drawing an arrow on each curve, going in the direction of increasing subscripts of the vertices. Now either $j$ preserves the orientation of both $A$ and $B$ or else $j$ reverses the orientation of both $A$ and $B$. In either case, we have $\mathrm{Lk}(A, B) = \mathrm{Lk}[j(A), j(B)]$. This leads to a contradiction because $j$ is orientation reversing and $\mathrm{Lk}(A, B) \neq 0$. Thus if $C_n$ has at least three rungs, then this embedding of $C_n$ is topologically chiral. Note that this same proof would also work to show that any closed ladder with at least three rungs that is embedded with any even number of half-twists is

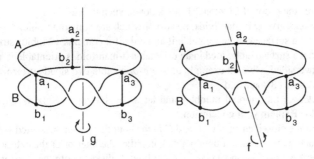

**Figure 3.41.** *g* rotates $C_n$ around a vertical axis while twisting, and *f* rotates $C_n$ around a horizontal axis

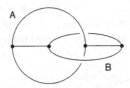

**Figure 3.42.** A two-rung cylinderical ladder with two half-twists is topologically achiral

topologically chiral, although from the previous argument we already knew this in the case in which there were at least three half-twists.

For $n = 1$ and $n = 2$, there is a topologically achiral embedding of $C_n$ as a strip with two half-twists. In particular, we can deform $C_2$ to the position illustrated in Figure 3.42, where the mirror plane contains the simple closed curve $A$ and the simple closed curve $B$ is embedded in a perpendicular plane. We can see from the illustration that this embedding of $C_2$ is its own mirror image. A topologically achiral embedding of $C_1$ is obtained by omitting one rung from $C_2$.

## Exercises

1. Draw a step-by-step deformation which takes the Mobius ladder in Figure 3.2 to its mirror image. Explain why an analogous deformation will not take the molecular Mobius ladder to its mirror image.
2. Explain in your own words how we know that triple-layered napthlenophane is topologically chiral.
3. Draw a topologically achiral $G(5, 4, 3)$.
4. Draw a topologically achiral $G(2, 5, 4)$.
5. Draw two different topologically achiral embeddings of $G(3, 3, 3)$.
6. In how many different ways can we choose the loop of $M_3$ in Figure 3.3? Explain.
7. The reflection illustrated in Figure 3.2 takes the loop $\overline{1234561}$ to what simple closed curve?

8. How does the proof of Lemma 3.1 break down when $n = 3$?
9. Draw at least five two-manifolds, no two of which are homeomorphic.
10. Prove that a surface is nonorientable if and only if it contains a Möbius strip.
11. Prove that a surface embedded in an orientable 3-manifold is orientable if and only if it is two sided. Give an example of a non-orientable two sided surface which is embedded in a 3-manifold.
12. Let $\mathbb{Q}$ denote the rational numbers and let $M = \{(x, y) \in \mathbb{R}^2 | x \in \mathbb{Q}\}$. Is $M$ a manifold? Explain your conclusion.
13. Draw simple closed curves $A$, $B$, and $C$ in the boundary of an unknotted solid torus in $\mathbb{R}^3$ such that $A$ does not bound a disk in either the interior or the exterior of the solid torus but does bound a disk in $\mathbb{R}^3$, $B$ bounds disks in both the interior and the exterior of the solid torus, and $C$ does not bound any disk in $\mathbb{R}^3$.
14. Consider $S^3 = \{(x, y, z, w) \in \mathbb{R}^4 | x^2 + y^2 + z^2 + w^2 = 1\}$. Define $p : S^3 \to \mathbb{R}^2$ by $p(x, y, z, w) = (x, y)$. What is the range of the map $p$? Let $A = \{(x, y) \in \mathbb{R}^2 | x^2 + y^2 \leqslant \frac{1}{2}\}$ and $B = \{(x, y) \in \mathbb{R}^2 | \frac{1}{2} \leqslant x^2 + y^2 \leqslant 1\}$. Describe what the sets $p^{-1}(A)$ and $p^{-1}(B)$ look like. Prove that $S^3 = p^{-1}(A) \cup p^{-1}(B)$.
15. Construct Seifert surfaces for the figure eight knot and the Hopf link, illustrated below.

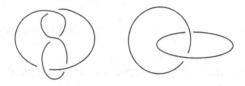

16. Consider a solid torus $V$ that is knotted as a figure eight knot. Draw $V$ with a longitude indicated.
17. Describe the three-manifold we obtain when we glue two solid tori together, meridian to meridian and longitude to longitude.
18. Construct the twofold branched cover of $S^3$ branched over a two-component unlink.
19. Find a deformation from a $M_3$ in the form of a Möbius strip to the embedding illustrated in Figure 3.23.
20. How would the proof of Theorem 3.2 break down if $n = 2$?
21. In the proof of Theorem 3.2, we obtained a contradiction from the assumption that $\overline{h}$ preserves the orientation of $\overline{\alpha 1}$. Prove that we get a similar contradiction if we assume that $\overline{h}$ reverses the orientation of $\overline{\alpha 1}$.

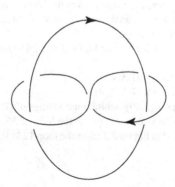

22. In the proof of Theorem 3.3, in Case 1, show that if we assume that $g$ reverses the orientation of $K_1$ then we get a contradiction. Also, in Case 2, show that if we assume that $g$ reverses the orientation of $K_i$ we get a contradiction.

23. What goes wrong if we try to adapt the proof of Theorem 3.3 to the graph of triple-layered naphthalenophane?

24. What goes wrong if we try to adapt the proof of Theorem 3.5 to the case in which not all of $p$, $q$, and $r$ are even?

25. Use the two different definitions of linking number to compute the linking number of the oriented link given below.

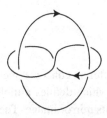

26. Suppose that $C_n$ is a closed circular ladder with $n$ rungs, embedded as a closed strip with at least two half-twists. Prove that every homeomorphism $h : (S^3, C_n) \rightarrow (S^3, C_n)$ must take rungs to rungs and sides to sides.

27. Consider a closed circular ladder $C_n$ embedded as a closed strip with $p$ half-twists. For what values of $p$ do the sides of $C_n$ form a knot and for what values do they form a link? Explain.

28. Let $x$ and $y$ be distinct points in $\mathbb{R}^n$. Prove that there are disjoint open sets $U$ and $V$ such that $x \in U$ and $y \in V$.

29. Suppose that $M$ is a twofold cover of $N$ with projection map $p$ and covering involution $h$. Prove that for every $x \in N$, there is an open set $U$ in $N$ with $x \in U$ such that $p^{-1}(U)$ is the union of two disjoint open sets $V_1$ and $V_2$ such that $p : V_1 \rightarrow U$ and $p : V_2 \rightarrow U$ are both homeomorphisms.

30. Construct the twofold branched cover of a two-dimensional sphere branched over four points.

31. Prove that the graph $G(p, q, p)$ has only three non-trivial automorphisms.

# 4

# Different Types of Chirality and Achirality

So far we have seen several different concepts of chirality and achirality. There is the experimental concept, which defines a molecule to be achiral if it can chemically change itself into its mirror image. There is the geometric concept, which defines a rigid graph to be achiral if it can be superimposed on its mirror image. Finally, there is the topological concept, which defines an embedded graph to be topologically achiral if it can be deformed to its mirror image. In this chapter, we introduce some additional approaches to achirality, which are topologically rigorous but try to capture some aspects of the rigidity of molecular structures.

Recall from Chapter 1 that the experimental concept of chirality does not necessarily coincide with the geometric concept. For example, we saw in Chapter 1 that Mislow's biphenyl derivative (Mislow & Bolstad, 1955) could chemically change itself into its mirror image, but its molecular bond graph could not be rigidly superimposed on its mirror image (see Figure 4.1).

A *symmetry presentation* of a graph is an embedding of the graph that can be rigidly superimposed on its mirror image. Building on a paper of Van Gulick which was originally written in 1960 and only published in 1993, Walba (1983) called a molecule a *Euclidean rubber glove* if it could chemically change itself into its mirror image but chemically it could not attain a symmetry presentation. This concept led him to ask: Could an embedded graph exist that could be deformed to its mirror image but that could not be deformed to a symmetry presentation? Walba conjectured that such graphs could exist and called any graph of this type a *topological rubber glove*. We will restate Walba's conjecture more formally by making use of the following definition.

**Definition.** A graph $G$ that is embedded in $S^3$ (or $\mathbb{R}^3$) is said to be *rigidly achiral* in $S^3$ (or $\mathbb{R}^3$) if there is an orientation-reversing finite-order homeomorphism of the pair $(S^3, G)$ [or the pair $(\mathbb{R}^3, G)$].

110

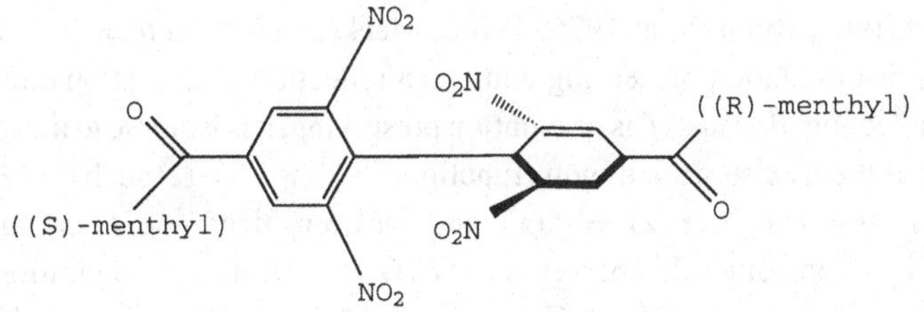

**Figure 4.1.** A molecule which can change itself into its mirror image but cannot be rotated to its mirror image

In order to help us understand this definition, we first consider an embedded graph $G$, which can be deformed to a symmetry presentation in $\mathbb{R}^3$. Thus there is an ambient isotopy $F : \mathbb{R}^3 \times I \to \mathbb{R}^3$ such that for each fixed $t \in I$ the function $F(x, t)$ is a homeomorphism, $F(x, 0) = x$ for all $x \in \mathbb{R}^3$, and $F(G \times \{1\}) = A$, where $A$ is a symmetry presentation of $G$. Also there is a rotation $r$ such that $A$ can be rotated to its mirror image. Because a graph has only finitely many vertices, this rotation must have finite order. Composing $r$ with a reflection gives us an orientation-reversing finite-order homeomorphism $h$ of the pair $(\mathbb{R}^3, A)$. Now let $f : \mathbb{R}^3 \to \mathbb{R}^3$ be given by $f(x) = F(x, 1)$. Then $f(G) = A$, and by conjugating $h$ by $f$ we obtain the homeomorphism $f^{-1} \circ h \circ f : (\mathbb{R}^3, G) \to (\mathbb{R}^3, G)$. Suppose that the order of $h$ is $n$. Now $(f^{-1} \circ h \circ f)^n = (f^{-1} \circ h \circ f) \circ (f^{-1} \circ h \circ f) \circ \cdots \cdot (f^{-1} \circ h \circ f) = f^{-1} \circ h \circ (f \circ f^{-1}) \circ h \circ (f \circ f^{-1}) \circ \cdots \cdot (f \circ f^{-1}) \circ h \circ f = f^{-1} \circ h^n \circ f = f^{-1} \circ f$, which is the identity function. Hence $f^{-1} \circ h \circ f$ also has finite order. Furthermore, $f$ is orientation preserving and $h$ is orientation reversing, so $f^{-1} \circ h \circ f$ is orientation reversing. Therefore, if $G$ can be deformed to a symmetry presentation, then there is an orientation-reversing finite-order homeomorphism of the pair $(\mathbb{R}^3, G)$. In other words, $G$ is rigidly achiral by our definition. For example, we can see that the graph of the molecule in Figure 4.1 is rigidly achiral by our definition as follows. First we deform the graph into a plane; then we reflect space through that plane (keeping every point in the plane fixed); then we undo the deformation so that the graph goes back to its original position. The final stage of the deformation is a homeomorphism $h$, taking the molecular graph to a graph in the plane. If we let $f$ be the reflection map, then $h^{-1} \circ f \circ h$ is an orientation-reversing homeomorphism of order two of $\mathbb{R}^3$ that takes the graph to itself.

Now we start with any graph $G$ that is rigidly achiral in $\mathbb{R}^3$. Then there is an orientation-reversing finite-order homeomorphism $h$ of the pair $(\mathbb{R}^3, G)$. A topological result that is not easy to prove is that any orientation-reversing finite-order homeomorphism of $\mathbb{R}^3$ is conjugate, by an orientation-preserving homeomorphism, to a reflection or to a rotation composed with a reflection

(Livesay, 1963; Rubinstein, 1976; Yau & Meeks, 1984). So $h = f^{-1} \circ r \circ f$ where $f$ is orientation preserving and $r$ is a reflection or a rotation composed with a reflection. Because $f$ is orientation preserving, it is isotopic to the identity map. Thus there exists a continuous function $F : \mathbb{R}^3 \times I \to \mathbb{R}^3$ such that $F(x, 0)$ is the identity map, $F(x, 1) = f(x)$, and for every fixed $t \in I$, the function $F(x, t)$ is a homeomorphism. Let $A = F(G \times \{1\})$; then by definition, $A$ is ambient isotopic to $G$. Also $f(G) = A$, so $f^{-1}(A) = G$ and $G = h(G) = (f^{-1} \circ r \circ f)(G)$. This means that $f(G) = r \circ f(G)$. If we substitute $A = f(G)$ into this last equation, we obtain $r(A) = A$. Thus $A$ is an embedded graph that is ambient isotopic to $G$ and such that there is a reflection or a rotation composed with a reflection $r$ that takes $A$ to itself. In other words, $A$ is a symmetry presentation to which $G$ can be deformed.

It follows from the above two paragraphs that a graph $G$ is rigidly achiral in $\mathbb{R}^3$ if and only if $G$ can be deformed to a symmetry presentation. Note that if an embedded graph is rigidly achiral, it is not necessarily rigidly superimposable on its mirror image. Our definition of rigid achirality permits a graph to be deformed to a symmetry presentation, whereas rigid superimposability implies that the graph is already embedded in a symmetry presentation. Rigid achirality is a topological concept, whereas rigid superimposability is a geometric concept.

Another observation we should make is that any orientation-reversing homeomorphism that has finite order necessarily has even order (the reader should prove this).

Unlike most of the concepts we have presented so far, the concept of rigid achirality in $\mathbb{R}^3$ is not the same as the concept of rigid achirality in $S^3$. A finite-order homeomorphism of $\mathbb{R}^3$ can be extended to a finite-order homeomorphism of $S^3$ by simply defining it to fix the point at infinity. Thus a graph that is rigidly achiral in $\mathbb{R}^3$ is necessarily rigidly achiral in $S^3$. In contrast, a finite-order orientation-reversing homeomorphism of $S^3$ may have no fixed points off of the graph, and in this case we would not be able to choose a point at infinity to omit such that $\mathbb{R}^3$ would still contain the entire graph and would be invariant under the finite-order homeomorphism. We illustrate this possibility with the figure eight knot in Figure 4.2. The two ends of the string are closed up at the point at infinity in $S^3$. Observe that this embedding is invariant under the orientation-reversing homeomorphism of $S^3$ given by a rotation by $180°$ about an axis going through the center point of the knot and the point at infinity, composed with a reflection through the plane of the paper. If we want to restrict a homeomorphism of $S^3$ to a homeomorphism of $\mathbb{R}^3$, we must select a point that is fixed by the homeomorphism to be our point at infinity in $S^3$. Both of the fixed points of the homeomorphism of Figure 4.2 are on the knot, so there is no way to restrict this homeomorphism of $S^3$ to a homeomorphism of the knot in $\mathbb{R}^3$.

**Figure 4.2.** In $S^3$, this knot can be rotated by 180° to obtain its mirror image

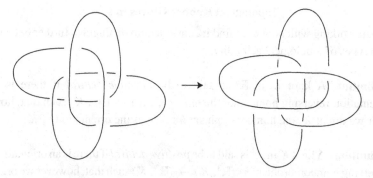

**Figure 4.3.** A symmetry presentation of the figure eight knot

In fact, in $\mathbb{R}^3$, there is no way to deform the figure eight knot to a position that can be rotated by 180° to get its mirror image (Flapan, 1987). In Chapter 1 we saw that the figure eight knot can be deformed to the symmetry presentation illustrated in Figure 4.3. However, this symmetry presentation must be rotated by 90° to obtain its mirror image. If we rotate the symmetry presentation by 180° we will obtain a projection identical to the one that we started with.

Returning to Walba's conjecture, we define a *topological rubber glove* to be an embedded graph that is topologically achiral but not rigidly achiral. In Chapter 2, we explained that an embedded graph is topologically achiral in $S^3$ if and only if it is topologically achiral in $\mathbb{R}^3$; also, we saw above that an embedded graph that is rigidly achiral in $\mathbb{R}^3$ must also be rigidly achiral in $S^3$, but not necessarily vice versa (we shall see examples of this shortly). It follows that a topological rubber glove in $S^3$ is a topological rubber glove in $\mathbb{R}^3$, but not necessarily vice versa.

In this chapter we give examples of several types of graphs that are topological rubber gloves. We start by finding a knot that is a topological rubber glove in $\mathbb{R}^3$ but is not a topological rubber glove in $S^3$. There do exist knots that are topological rubber gloves in $S^3$, but they are very complicated to draw and to explain, so we will not present them here (see Flapan, 1987). However, we do present three different graphs that are topological rubber gloves in $S^3$ and hence in $\mathbb{R}^3$ as well. Two of these graphs are molecular graphs, and the third is an intrinsic topological rubber glove in the sense that, although there exists an embedding of the graph that is topologically achiral, there does not exist an embedding of the graph that is rigidly achiral. Finally, we will present an alternative approach to molecular chirality by using cell complexes to represent molecules rather than representing molecules with molecular graphs.

## Topological Rubber Gloves in $\mathbb{R}^3$

When working with knots, we find it convenient to distinguish further between two types of topological achirality.

**Definition.** A knot $K$ in $\mathbb{R}^3$ is said to be *negative achiral* if there is an orientation-reversing homeomorphism $h : (\mathbb{R}^3, K) \to (\mathbb{R}^3, K)$ such that, however we orient $K$, the homeomorphism $h$ reverses the orientation of $K$.

**Definition.** A knot $K$ in $\mathbb{R}^3$ is said to be *positive achiral* if there is an orientation-reversing homeomorphism $h : (\mathbb{R}^3, K) \to (\mathbb{R}^3, K)$ such that, however we orient $K$, then $h$ preserves the orientation of $K$.

By replacing $\mathbb{R}^3$ in the above definitions by $S^3$, we obtain analogous definitions in $S^3$. As with topological achirality, a knot is negative or positive achiral in $\mathbb{R}^3$ if and only if it is correspondingly negative or positive achiral in $S^3$. Any topologically achiral knot must be either positive achiral, negative achiral, or both. Figure 4.4 illustrates a knot in $\mathbb{R}^3$ that is both negative and positive achiral. We see that this knot is negative achiral because it has a vertical planar reflection that takes the knot to itself and reverses the orientation of the knot. We see that it is positive achiral because we can rotate it by 180° about a central axis, perpendicular to the plane of the paper, to get the mirror image of the knot. This rotation preserves the orientation on $K$.

We say that a knot is *rigidly negative achiral* or *rigidly positive achiral* in $\mathbb{R}^3$ or $S^3$ if the homeomorphisms in the above definitions have finite order. Any rigidly achiral knot must be either rigidly negative achiral or rigidly positive achiral. We can see from Figure 4.2 that the figure eight knot is rigidly negative achiral in $S^3$, because the homeomorphism of $S^3$ illustrated in Figure 4.2 reverses the orientation on the knot. We can deform this homeomorphism of $S^3$ to a homeomorphism of $\mathbb{R}^3$, which still reverses the orientation on the knot, by holding some point off of the knot fixed. However, doing this will cause the homeomorphism to no longer be of finite order. Thus, the figure eight knot will also be negative achiral in $\mathbb{R}^3$, but not necessarily rigidly negative achiral. We can see that the knot in Figure 4.4 is both rigidly negative achiral and rigidly positive achiral in $\mathbb{R}^3$. In fact, we shall prove that any rigidly negative achiral knot in $\mathbb{R}^3$ looks essentially like the example in Figure 4.4. First we need the following definition.

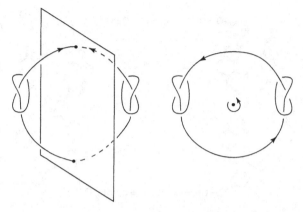

**Figure 4.4.** This knot is both negative achiral and positive achiral in $\mathbb{R}^3$

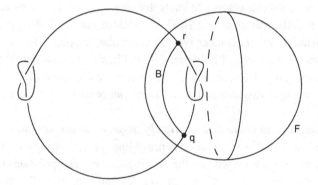

**Figure 4.5.** This knot is a connected sum

**Definition.** Let $K$ be a knot in $S^3$ or $\mathbb{R}^3$. Suppose that there exists a surface $F$ homeomorphic to a two-dimensional sphere such that $F$ meets $K$ in two points $p$ and $q$. Let $B$ be an arc in $F$ with endpoints $p$ and $q$. Consider the simple closed curves obtained by joining each of the components of $K - \{p, q\}$ to the arc $B$. If $F$ can be chosen so that neither of these simple closed curves is the unknot, then we say that $K$ is a *composite knot* or a *connected sum.*

The knot in Figure 4.4 is a connected sum, and Figure 4.5 illustrates the sphere $F$ that splits this knot into two knotted arcs.

In 1996, Carina, Dietrich-Buchecker, and Sauvage synthesized the first composite knots (Carina et al., 1996). We illustrate one such knotted molecule in Figure 4.6. These molecular composite knots are made out of two trefoil knots.

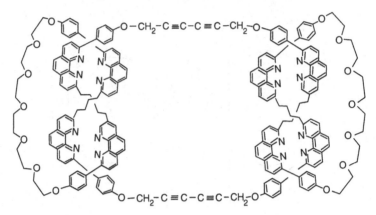

**Figure 4.6.** Two molecular composite knots

Some of the molecular composite knots that they synthesized consisted of two identical trefoils, and some consisted of a right-handed trefoil together with a left-handed trefoil. This latter type of molecular composite knots have the form of the knot illustrated in Figure 4.4, and hence are both rigidly positive achiral and rigidly negative achiral. In contrast, those molecular composite knots consisting of two identical trefoil knots can be shown to be topologically chiral.

We would like to prove that any rigidly negative achiral knot in $\mathbb{R}^3$ is necessarily a connected sum. In order to prove this, as well as most of the other results in this chapter, we will use the classification of fixed-point sets of finite-order homeomorphisms of $S^3$, which was developed by P. A. Smith in 1939 (Smith, 1939). In fact, Smith's work went way beyond this, to consider finite-order homeomorphisms of homology spheres of various dimensions, but for our purposes we are only interested in homeomorphisms of $S^3$ and $\mathbb{R}^3$. The *fixed-point set* of a homeomorphism $h : S^3 \to S^3$ is the set of all points $x$ such that $h(x) = x$. Smith's Theorem 4.1 states the part of Smith's work that we shall use. The proof of this theorem is beyond the scope of this text.

**Smith's Theorem 4.1.** *Let* h $: S^3 \to S^3$ *be a finite-order homeomorphism. If* h *is orientation preserving, then the fixed-point set of* h *is either the empty set or is homeomorphic to a circle. If* h *is orientation reversing, then the fixed point set of* h *is either two points or is homeomorphic to a two-dimensional sphere.*

We shall also need to understand the fixed-point sets of homeomorphisms of $\mathbb{R}^3$. We can figure this out easily from Smith's Theorem 4.1 as follows. We start with a finite-order homeomorphism of $h : \mathbb{R}^3 \to \mathbb{R}^3$, which we extend to

$S^3$ by defining $h(\infty) = \infty$. Now $h : S^3 \to S^3$ is a finite-order homeomorphism that will be orientation preserving if and only if $h : \mathbb{R}^3 \to \mathbb{R}^3$ was orientation preserving. If $h : \mathbb{R}^3 \to \mathbb{R}^3$ is orientation preserving, then the fixed-point set of $h : S^3 \to S^3$ is homeomorphic to a circle, because the fixed-point set of $h$ necessarily contains the point $\infty$ and so cannot be the empty set. It follows that the fixed-point set of $h : \mathbb{R}^3 \to \mathbb{R}^3$ is homeomorphic to a line, because if we remove the point at infinity from a set homeomorphic to a circle, then we obtain a set homeomorphic to a line. If $h$ is orientation reversing then the fixed-point set of $h : S^3 \to S^3$ is either homeomorphic to a two-dimensional sphere or is two points. In either case, the fixed-point set of $h$ contains the point $\infty$. Because a two-dimensional sphere with a point removed is homeomorphic to a plane, the fixed-point set of $h : \mathbb{R}^3 \to \mathbb{R}^3$ is either homeomorphic to a plane or is a single point. With Smith's Theorem 4.1 in hand, we can now prove the following theorem (Flapan, 1987).

**Theorem 4.2.** *If a nontrivial knot is rigidly negative achiral in* $\mathbb{R}^3$, *then it is a composite knot.*

*Proof.* Suppose that $K$ is a nontrivial knot that is rigidly negative achiral in $\mathbb{R}^3$. Pick an orientation for $K$ and let $h : (\mathbb{R}^3, K) \to (\mathbb{R}^3, K)$ be a finite-order orientation-reversing homeomorphism that reverses the orientation on $K$. Because $K$ is homeomorphic to a circle and $h$ has finite order, $h$ either reflects $K$ or rotates $K$. We know that $h$ reverses the orientation on $K$, so $h$ must reflect $K$. Because $h$ has finite order, this means that $h$ fixes two points on $K$, say $p$ and $q$. We can extend $h$ to a homeomorphism of $(S^3, K)$ by defining $h(\infty) = \infty$. Then $h : (S^3, K) \to (S^3, K)$ will still be an orientation-reversing finite-order homeomorphism. Because $K$ was contained in $\mathbb{R}^3$ the point $\infty$ is not on $K$. So $h : (S^3, K) \to (S^3, K)$ has at least three fixed points, $p, q,$ and $\infty$. Thus by Smith's Theorem 4.1, the fixed point set of $h$ must be a set $F$ that is homeomorphic to a two-dimensional sphere. Because $h$ is orientation reversing and $h$ fixes each point of $F$, it must exchange the two components of $S^3 - F$. The fixed-point set $F$ contains the points $p$ and $q$ and no other points of $K$. Let $B$ be any arc in $F$ with endpoints $p$ and $q$, and let $K_1$ and $K_2$ be the simple closed curves obtained by joining each of the components of $K - \{p, q\}$ to the arc $B$. The homeomorphism $h$ interchanges the two components of $K - \{p, q\}$ and fixes every point of $B$. Thus $h$ exchanges $K_1$ and $K_2$; so either both $K_1$ and $K_2$ are knotted (and one knot is the mirror image of the other) or neither is knotted. If neither were knotted, then $K$ would be the unknot, contrary to hypothesis. Hence both $K_1$ and $K_2$ must be knotted. Therefore by definition, $K$ is a composite knot. $\square$

**Figure 4.7.** The knot $8_{17}$ is not a composite knot and it is negative achiral but not positive achiral in $S^3$

**Figure 4.8.** A deformation of $8_{17}$ to its mirror image, which reverses the orientation on the knot

Theorem 4.2 helps us to construct examples of knots that are topological rubber gloves in $\mathbb{R}^3$. We pick a knot that is not a composite knot and that is negative achiral but not positive achiral. The knot with the fewest crossings that has this property is the knot $8_{17}$, which is illustrated in Figure 4.7.

Kawauchi (1979) proved that $8_{17}$ is not positive achiral in $S^3$. We can see that this knot is negative achiral in $\mathbb{R}^3$ by the deformation of it to its mirror image illustrated in Figure 4.8. We rotate the knot by 180°, and then we pull the long arc from the bottom of the knot back to the top of the knot. We then follow this deformation by a mirror reflection through the plane of the paper to get our original projection of $8_{17}$, with its orientation reversed. Thus the knot is negative achiral in $\mathbb{R}^3$, and hence it is topologically achiral in $\mathbb{R}^3$.

Because $8_{17}$ is not a composite knot it cannot be rigidly negative achiral in $\mathbb{R}^3$ by Theorem 4.2. Because it is not positive achiral in $S^3$, it is also not positive achiral in $\mathbb{R}^3$; thus it is certainly not rigidly positive achiral in $\mathbb{R}^3$. Hence $8_{17}$ is topologically achiral but not rigidly achiral in $\mathbb{R}^3$. Therefore $8_{17}$ is a topological rubber glove in $\mathbb{R}^3$.

However, $8_{17}$ is not a topological rubber glove in $S^3$ because it is rigidly negative achiral in $S^3$. To see this we deform $8_{17}$ in $S^3$ so that the point at infinity is on the long arc of $8_{17}$. Then, to get its mirror image, we can rotate the knot by 180° about an axis perpendicular to the page that meets the knot at its center point and at the point at infinity. Taking the mirror image after rotating by 180° gives us $8_{17}$ with its orientation reversed. This symmetry presentation for $8_{17}$ in $S^3$ is illustrated in Figure 4.9, with the point where the axis meets the knot indicated.

### Molecular Topological Rubber Gloves

Although the knot $8_{17}$ is a topological rubber glove in $\mathbb{R}^3$, it is not a molecular bond graph. The knot $8_{17}$ is more complicated than any of the molecular knots

**Figure 4.9.** A symmetry presentation for $8_{17}$ in $S^3$

that have been synthesized thus far. For many years after it was known that knots could be topological rubber gloves, it was still unknown whether a molecular structure could be a topological rubber glove. In 1992, Du and Seeman synthesized a figure eight knot from single-stranded DNA (Du & Seeman, 1992), which was later shown to be a topological rubber glove (Flapan & Seeman, 1995). The global structure of this synthetic single-stranded DNA molecule is a figure eight knot, but locally the molecular graph is more complicated. The global structure of the simple closed curve making up the DNA is known as the *backbone* of the DNA. Locally, the graph is made up of 66 to 104 *nucleotide subunits*, which each contain a phosphate, a sugar, and one of the four bases cytosine, thymine, adenine, and guanine. Figure 4.10A is an illustration of the molecular graphs of the four bases and Figure 4.10B is an illustration of two nucleotide subunits. The letter $R$ indicates the location where the bases are attached to the sugars. (For a more detailed discussion of the topology of DNA, see Chapter 7.)

We prove that the DNA figure eight knot is topologically achiral as follows. First we deform the figure eight backbone to the symmetry presentation of the figure eight knot illustrated in Figure 4.11. If we rotate this presentation of the figure eight knot by 90°, we will obtain its mirror image which can now be deformed to the mirror image of the original embedding. Rotating the backbone of the molecular graph has the unwanted side effect of moving the bases to new positions relative to the knot. However, we can now slither the structure along itself until each base is back to its original position. We then deform the sugars and bases so that each will be in the mirror-image position of where it was to start. Thus the molecular bond graph of the DNA figure eight is topologically achiral.

We now prove that the molecular graph $G$ of the DNA figure eight is not rigidly achiral in $S^3$ and hence not in $\mathbb{R}^3$ either. Suppose that there exists an orientation-reversing finite-order homeomorphism $h : (S^3, G) \rightarrow (S^3, G)$. In particular, $h$ must take each base to a base of the same type, as the bases cytosine, thymine, adenine, and guanine all have distinct molecular graphs. However, the bases in the synthetic DNA figure eight do not occur in a sequence that repeats itself some number of times, and they do not occur in a sequence that is palindromic (i.e., the same when read backward or forward). Hence $h$

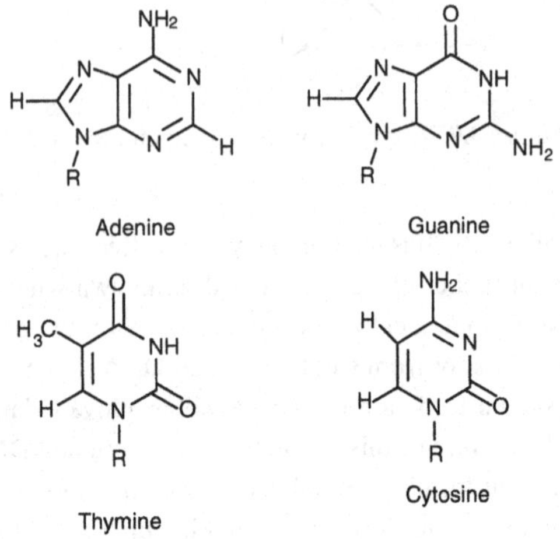

Adenine

Guanine

Thymine

Cytosine

**Figure 4.10A.** The bases cytosine, thymine, adenine, and guanine

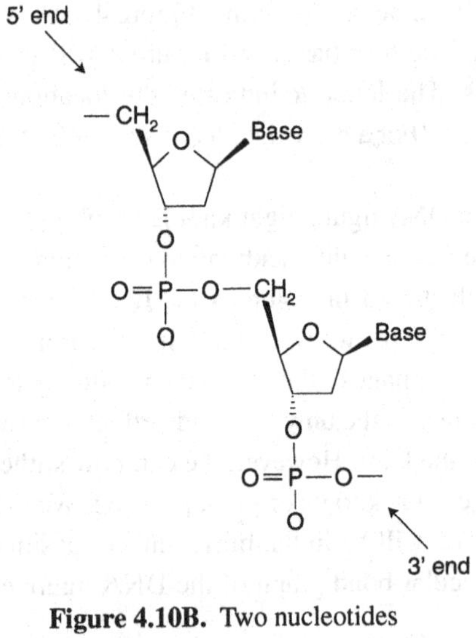

5' end

— CH₂        Base

O=P—O—CH₂        Base

O=P—O —

3' end

**Figure 4.10B.** Two nucleotides

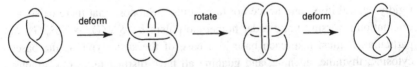

**Figure 4.11.** A figure eight knot can be deformed to a symmetry presentation, which can be rotated to its mirror image then deformed to the mirror image of the original embedding

can neither rotate each base to a similar base along the backbone nor reflect the backbone in such a way that each base is interchanged with a base of the same type. So $h$ must send each base to itself. We can see from the molecular graphs drawn in Figures 4.10A and B that this means that $h$ must fix every point of the entire molecular graph, except possibly for some hydrogen atoms. But by Smith's Theorem 4.1, the fixed point set of $h$ is either two points or is homeomorphic to a two-dimensional sphere. In either case, the molecular graph (possibly with some hydrogen atoms removed) must be contained in a set that is homeomorphic to a two-dimensional sphere. However, because the graph has the global structure of a nontrivial knot, it cannot lie in the surface of a plane. A two-dimensional sphere is homeomorphic to a plane together with a point at infinity. If a graph $G$ could be contained in a two-dimensional sphere $S$, then we could choose the point at infinity to be in $S - G$, so that $G$ could also be contained in a plane. Thus it is impossible for the graph $G$ to be contained in a set homeomorphic to a two-dimensional sphere. Hence no such homeomorphism can exist. Therefore, the synthetic single-stranded DNA figure eight knot is a topological rubber glove in both $S^3$ and $\mathbb{R}^3$.

Although this is the first molecular topological rubber glove known to exist, it is not chemically achiral. In particular, although the graphs of the sugars can be deformed to their mirror images, on a chemical level, because of rigidity, sugars cannot convert themselves to their mirror images. Thus chemically this molecule is chiral. In contrast, Chambron et al. (1997) have recently designed and synthesized an entirely new kind of molecular topological rubber glove. Their molecule is a nontrivial link that is chemically achiral but not rigidly achiral. Because chemical achirality implies topological achirality, their molecule is a topological rubber glove. This molecule is noteworthy, both because it is the first molecular topological rubber glove that is not made from DNA and because it is the first topological rubber glove that is actually chemically achiral. We illustrate the molecule of Chambron, et al. in Figure 4.12.

We can see how to deform this molecule to its mirror image as follows. Let $G$ denote the molecular graph that is illustrated in Figure 4.12. In order to obtain the mirror image of $G$ we turn the lower ring over so that the $H_3C$ moves to the right-hand side, while rotating the naphthalene (i.e., the pair of benzene rings that are fused together) at the top of the graph by $180°$. In this way, we get the mirror image of $G$ through a vertical plane that is perpendicular to the plane of the paper. Note that we write the $H_3C$ on the right side as $CH_3$ because we want to make it clear that it is the carbon that is bonded with the benzene ring. These two deformations can both occur chemically, so the molecular link is chemically achiral as well as topologically achiral.

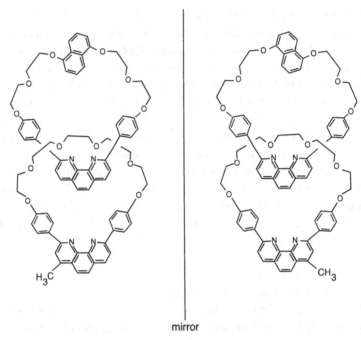

mirror

**Figure 4.12.** A topological rubber glove that is chemically achiral

Now we shall prove that $G$ cannot be rigidly achiral. Let $A$ denote the component of the link containing the $H_3C$, and let $B$ denote the other component of the link. Suppose that $G$ is rigidly achiral in $S^3$. Then there is a finite-order orientation-reversing homeomorphism $h : (S^3, G) \rightarrow (S^3, G)$. Because $h$ is orientation reversing, by Smith's Theorem 4.1, the fixed point set of $h$ is either two points or a set that is homeomorphic to a two-dimensional sphere. There is only one $H_3C$ in the graph, so $h$ must pointwise fix this $H_3C$, together with the edge connecting $H_3C$ to the rest of $A$. This means that $h$ fixes more than two points of $S^3$, so the fixed point set of $h$ must be a set $P$ that is homeomorphic to a two-dimensional sphere.

Next we consider the component $B$ of the graph. As there is only one naphthalene in the graph, $h$ must take the naphthalene to itself, either exchanging the two hexagons or taking each hexagon to itself. Let $e$ denote the edge along which the two hexagons are fused. This particular edge is unique in the graph, so $h(e) = e$. This means that either $e$ is fixed pointwise by $h$, or $h$ flips $e$ over, fixing a single point of $e$. Suppose that $h$ flips $e$ over, fixing a single point of $e$. Then $e$ intersects the fixed point set, $P$, in precisely one point, so each of the hexagons in the naphthalene intersects the set $P$. Because $P$ separates $S^3$ into two components, every simple closed curve that intersects $P$ must intersect $P$ in an even number of points. So both of the hexagons in the naphthalene must intersect $P$ in at least one point that is not on the edge $e$. Thus $h$ fixes an additional point of each hexagon in the naphthalene. Hence $h$ cannot exchange the two benzene rings. It follows that the two points, say $x$ and $y$, connecting the naphthalene to the rest of $B$ must each be fixed, rather than exchanged. However, there is no way this can occur if $e$ is flipped over because one end of $e$ is closer than the other end of $e$ to $x$. Hence $h$ could not actually flip $e$ over, so $h$ must fix $e$ pointwise. Now $h$ cannot exchange the two hexagons in the naphthalene because $x$ is closer than $y$ is to one end of $e$. So $h$ must take each hexagon to itself. Because $h$ has an order of two, it follows that $h$ must pointwise fix the naphthalene. Hence $h$ pointwise fixes the chains of oxygens and carbons connecting the naphthalene to the hexagons on each side. Now $h$ takes each of these isolated hexagons to itself setwise, although they are not necessarily fixed pointwise by $h$. However, on each of these hexagons, $h$ must pointwise fix the two vertices connecting the benzene to the rest of $B$. Thus $h$ also pointwise fixes the *diazaphenanthrene* (that is, the group of three fused hexagons) on $B$, as well as the edges connecting the diazaphenanthrene to the nearby hexagons. Thus $h$ must pointwise fix every point on $B$ except possibly those on the isolated hexagons on either side of $B$.

Now we consider the component $A$ of the graph. Because $h$ fixes the $H_3C$ and the edge connecting it to the rest of $A$, the homeomorphism $h$ must also pointwise fix the diazaphenanthrene on $A$ as well as an edge on each side connecting it to the nearby hexagons. As we saw above for component $B$, the homeomorphism $h$ takes each of these two hexagons to itself setwise, and it pointwise fixes the two vertices connecting each hexagon to the rest of $A$. However, $h$ does not necessarily fix these two hexagons pointwise. Nonetheless, $h$ must also pointwise fix the chain of oxygens and carbons joining these two hexagons across the top of $A$.

Thus we have seen so far that all of the graph $G$ except possibly for two hexagons on each component must be fixed pointwise by $h$, and hence must be contained in the fixed point set $P$. We know that the vertices where each of these four hexagons is connected to the rest of the graph are each contained in $P$, and $h$ takes each of these four isolated hexagons to itself setwise. Because $h$ induces an order-one or order-two homeomorphism on each of these hexagons, either $h$ fixes the hexagon pointwise or $h$ exchanges the two sides of the hexagon. Suppose that at least one of the hexagons is not pointwise fixed by $h$. The four hexagons bound disks in the plane of the paper that we can see are disjoint from each other and from the rest of the graph. Let us consider such a disk for a hexagon that is not pointwise fixed by $h$. We shall denote this disk by $D$. Because $h$ interchanges the two sides of the hexagon, we know that the hexagon bounding $D$ has one arc on one side of $P$ and the other arc on the other side of

$P$, and the two arcs intersect at the two vertices of $D$ that are contained in $P$. Thus there must be some arc of intersection of $D$ and $P$ that joins these two points. As this arc is contained in $D$, its interior does not intersect the graph $G$ or any of the other disjoint planar disks bounded by hexagons. So, for the set of hexagons that are not pointwise fixed by $h$, we obtain disjoint arcs in $P$, which are also disjoint from the rest of $G$.

Now we create a new graph $H$ by replacing the hexagons that are not pointwise fixed by $h$ by the arcs described above. Thus the graph $H$ is entirely contained in the set $P$. However, $H$ has a subgraph that is a Hopf link, and yet $P$ is homeomorphic to a two-dimensional sphere. As a Hopf link is nontrivial it cannot be embedded in a two-dimensional sphere. So we obtain a contradiction. Therefore, there can be no finite-order orientation-reversing homeomorphism of $(S^3, G)$, so $G$ cannot be rigidly achiral in $S^3$ or in $\mathbb{R}^3$. Thus, the molecule illustrated in Figure 4.12 is a chemically achiral topological rubber glove.

## Intrinsic Topological Rubber Gloves

In all of the examples of topological rubber gloves that we have given so far, the specific embeddings of the graphs prevent them from being rigidly achiral. In particular, if the graphs are re-embedded so that they lie in the plane, then they will become rigidly achiral. We saw in Chapter 3 that the Möbius ladders with an odd number of rungs are intrinsically chiral. It is natural to ask whether there are topological rubber gloves that are intrinsic in the sense that, while there exists an embedding of the graph that is topologically achiral, there does not exist an embedding of the graph that is rigidly achiral. Observe as follows that there cannot exist a graph that is an intrinsic topological rubber glove in the sense that every embedding of it is a topological rubber glove. On one hand, any graph that contains a simple closed curve has knotted embeddings that are topologically chiral, and hence these embeddings are not topological rubber gloves. On the other hand, any graph that does not contain a simple closed curve has a planar embedding that is rigidly achiral, and hence is not a topological rubber glove.

Now we will construct an embedded graph that is topologically achiral, and yet there does not exist any embedding of the graph in $S^3$ that is rigidly achiral. The embedded graph $G$, which we consider, is made up of a Möbius ladder with four rungs $M$ that is embedded in $S^3$ so that it is topologically achiral, and is joined to a planar triangle $T$ by two edges $e_1$ and $e_2$, as illustrated in Figure 4.13. An orientation-reversing homeomorphism $h : (S^3, G) \to (S^3, G)$ of this embedding is obtained by rotating the Möbius ladder by $90°$ while fixing $T$, $e_1$, and $e_2$ and reflecting the entire graph through a horizontal plane containing the vertex $v$. Thus, this embedding of $G$ is topologically achiral.

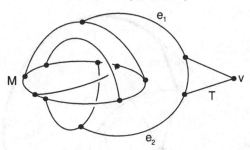

**Figure 4.13.** An embedded graph that is topologically achiral but does not have an embedding that is rigidly achiral

Now suppose that our graph $G$ is embedded in $S^3$ in such a way that there exists a finite-order orientation-reversing homeomorphism $h : (S^3, G) \to (S^3, G)$. Because of the structure of the graph, we must have $h(M) = M$, $h(T) = T$, and $h(v) = v$. Also $h$ must induce an order-one or order-two automorphism on the vertices of $T$, because these are the only automorphisms that $T$ has. In either case, $h^2$ must fix every point of $T$, and hence every point of the edges $e_1$ and $e_2$ that connect $T$ to $M$. We know that $h^2$ has finite order. So by Smith's Theorem 4.1, because $h^2$ must be orientation preserving, if it is not the identity map then its fixed-point set is either the empty set or a set homeomorphic to a circle. However, $h$ fixes every point of $T$ as well as the edges $e_1$ and $e_2$, so the fixed-point set of $h$ cannot be either the empty set or a set homeomorphic to a circle. Therefore $h^2$ must be the identity map. However, we can now apply the results of Chapter 3. Because $M$ is a Möbius ladder with four rungs, we know by Lemma 3.1 that if $K$ is the loop of $M$ then $h(K) = K$. Now because $h$ has an order of two, we must have $h(\alpha_i) = \alpha_i$, for each rung $\alpha_i$. This contradicts Theorem 3.3; hence no such homeomorphism $h$ can exist. Thus no embedding of $G$ in $S^3$ or $\mathbb{R}^3$ is rigidly achiral, so $G$ is an intrinsic topological rubber glove.

## Hierarchies of Achirality

We have seen that, in a sense, some graphs have embeddings which are more achiral than others in $\mathbb{R}^3$ or $S^3$. We can organize all abstract graphs into a hierarchy of how achiral their embeddings can be in $S^3$. We shall consider rigid achirality to be a stronger type of achirality than topological achirality in the sense that any embedded graph that is rigidly achiral is necessarily topologically achiral, but not vice versa. Hierarchies of chirality for embedded graphs have been developed by various people (see, e.g., Liang & Mislow, 1994a; Simon,

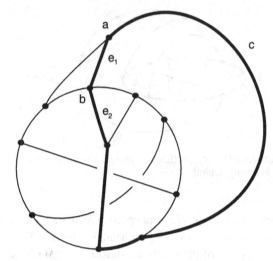

**Figure 4.14.** A graph that has a rigidly achiral embedding in $S^3$ but has no rigidly achiral embedding in $\mathbb{R}^3$

1987; Walba, 1985); however, the following is a hierarchy of the existence of achiral embeddings for abstract graphs.

1. Graphs that are intrinsically chiral.

   *Example.* A Möbius ladder with five rungs.

2. Graphs that can be embedded in $S^3$ so that they are topologically achiral, but cannot be embedded so that they are rigidly achiral.

   *Example.* The graph given in Figure 4.13.

3. Graphs that can be embedded in $S^3$ so that they are rigidly achiral.

   *Example.* Any graph that has a planar embedding.

We have an analogous hierarchy with respect to embeddings of graphs in $\mathbb{R}^3$ with the same examples as those given above. Note that the set of graphs of type 1 is the same in $\mathbb{R}^3$ and $S^3$, and any graph that is of type 3 in $\mathbb{R}^3$ is also of type 3 in $S^3$. However, there exist graphs that are of type 3 in $S^3$, yet are of type 2 in $\mathbb{R}^3$. For example, consider the graph $G$ embedded in $\mathbb{R}^3$ as illustrated in Figure 4.14. This graph is a Möbius ladder $M_4$ together with the additional edges $e_1$ and $e_2$, as well as their vertices. Suppose there is an embedding of $G$ in $\mathbb{R}^3$ such that there exists a finite-order orientation-reversing homeomorphism

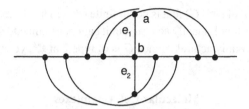

**Figure 4.15.** A rigidly achiral embedding of the graph $G$ in $S^3$

$h : (\mathbb{R}^3, G) \to (\mathbb{R}^3, G)$. Because the vertex $b$ is the only one in the graph with a valence of four, we must have $h(b) = b$. There is a unique simple closed curve $C$ in the graph that contains vertex $b$ and contains precisely five edges. The curve $C$ is darkened in Figure 4.14. Therefore, because $h$ is a homeomorphism and $h(b) = b$, we must have $h(C) = C$. From this it follows that $h$ leaves $e_1 \cup e_2$ setwise invariant. By omitting the edges $e_1$ and $e_2$, we obtain a four-rung Möbius ladder $M_4$ that must also be invariant under $h$. Let $K$ denote the loop of $M_4$. By Lemma 3.1, $h(K) = K$. Furthermore, because $K$ is unique, it must be the round circle in the picture. Because $h$ has finite order and $h$ fixes the vertex $b$ of $K$, either $h$ fixes every point of $K$ or there must be one other point of $K$ that is fixed by $h$. In either case, because $h$ is orientation reversing and fixes more than one point of $\mathbb{R}^3$, the fixed-point set of $h$ must be homeomorphic to a plane. If every point of $K$ were fixed by $h$, then every point of $G$ would be fixed by $h$. This is impossible because $G$ contains a $K_{3,3}$ graph and hence cannot be embedded in a plane. Thus precisely two points of $K$ are fixed by $h$. This means that the plane of fixed points must intersect $K$ in the point $b$ and one other point, in such a way that there are the same number of vertices on each side of the plane. Hence each rung of $M_4$ has one vertex on either side of the plane. So each rung $\alpha_i$ must intersect the plane of fixed points, and hence $h$ fixes at least one point of each rung. This implies that $h(\alpha_i) = \alpha_i$ for each rung $\alpha_i$, which contradicts Theorem 3.3. Therefore there is no rigidly achiral embedding of the graph $G$ in $\mathbb{R}^3$.

In contrast, if $h$ is a homeomorphism of $S^3$, it could fix two points of $K$ without necessarily fixing a plane of points, so the above proof does not work in $S^3$. In particular, in Figure 4.15 we have illustrated a rigidly achiral embedding of the graph $G$ in $S^3$. Here $K$ looks like a line because $K$ goes through the point at $\infty$. The finite-order orientation-reversing homeomorphism of $(S^3, G)$ rotates $G$ by $180°$ about an axis containing the vertex $b$ while reflecting $G$ through the plane of the paper. By deforming this homeomorphism slightly to fix a new point at infinity, which is off of the graph, we obtain an orientation-reversing homeomorphism of $(\mathbb{R}^3, G)$ that is not of finite order. Thus $G$ can be embedded in $\mathbb{R}^3$ so that it is topologically achiral, but it cannot be embedded in $\mathbb{R}^3$ in such a way that it is rigidly achiral. Hence $G$ is of type 2 in $\mathbb{R}^3$, yet of type 3 in $S^3$.

## Molecular Cell Complexes

Although the concept of rigid achirality characterizes the notion of when a molecular graph can be deformed to a symmetry presentation, such a deformation is often not chemically achievable because different bonds will flex and twist different amounts. As a topological object, an embedded graph omits information about which deformations are not chemically achievable. In this section we suggest an alternative approach to analyzing the topological chirality of complex molecules by modeling these molecules with cell complexes whose topology contains at least some of this information.

In particular, from a chemical point of view, not all rings behave as ordinary simple closed curves in an embedded graph. As a topological object, a simple closed curve may contain a knot or be linked with another simple closed curve, and a deformation of an embedded graph may even push a piece of the graph through the center of a simple closed curve. Although some large molecular rings have a lot of flexibility and can be embedded in different ways in space, not all molecular rings behave in this way. In particular, small or rigid molecular rings, as well as molecular rings with rigid pieces of the graph protruding into their center, may be impenetrable (i.e., nothing can pass through them). The chemical structure of such rings does not permit them to be knotted, linked, or have a piece of the graph pushed through the center. For example, benzene rings and other small rings have this property, though they are certainly not the only rings that are impenetrable. Throughout the remainder of this chapter, we shall use the term *impenetrable ring* to refer to any molecular ring through which no other piece of the molecule can pass. Not only does this mean that no piece of the graph can pass through that ring during a deformation; it means that even for a different embedding of the graph (i.e., a stereoisomer of the molecule), a piece of the graph cannot pierce through that ring. One approach to modeling the topological properties of molecular structures is to represent such impenetrable rings as polygonal surfaces, rather than as polygonal curves. If it is not geometrically apparent which rings are impenetrable, then at the very least, we can represent all the small rings, that is the 5, 6, and 7 vertex rings, as polygonal surfaces, as the small size of such rings forces then to be impenetrable. Any surface that is homeomorphic to a disk is said to be a *cell*, or more specifically a *two-dimensional cell*. Thus we shall refer to these polygonal surfaces as cells. In fact, any chain of two or

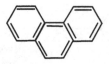

**Figure 4.16.** A chain of fused rings can be considered as a single cell

more polygonal surfaces that are sequentially fused together along edges is still homeomorphic to a disk, and so can be regarded as a single cell. Figure 4.16 illustrates a chain of three fused rings that can be considered as either three adjacent cells or as a single cell, depending on which is most convenient for a particular proof.

We shall refer to an embedded graph that has cells attached to each of the impenetrable rings as a *molecular cell complex*, or, more generally, an *embedded cell complex*. By analogy with our definition of topologically chiral embedded graphs, we have the following definition.

**Definition.** A cell complex $G$ embedded in $\mathbb{R}^3$ is *topologically achiral* if there is an orientation-reversing homeomorphism of $(\mathbb{R}^3, G)$. If no such homeomorphism exists, then we say that $G$ is *topologically chiral*.

This definition of topological achirality of cell complexes is equivalent to defining an embedded cell complex to be topologically achiral if it can be deformed to its mirror image. On a chemical level, no part of the molecule can pass through an impenetrable ring, so if a molecular cell complex is topologically chiral, then the molecule it represents will necessarily be chemically chiral. We shall use the following result, which we do not prove, but whose proof is fairly straightforward.

**Theorem 4.3.** *Let G be an embedded cell complex, and let* $C_1, C_2, \ldots, C_n$ *be some collection of cells in G. For each* $i = 1, 2, \ldots, n$, *let* $x_i$ *and* $y_i$ *be points on the boundary of* $C_i$ *and let* $A_i$ *be a non–self-intersecting arc in* $C_i$ *with endpoints* $x_i$ *and* $y_i$. *Suppose there is an orientation-reversing homeomorphism of* $(\mathbb{R}^3, G)$, *that takes the collection of cells* $C_1, C_2, \ldots, C_n$ *to itself (possibly permuting them), and takes the set of points* $\{x_1, y_1, x_2, y_2, \ldots, x_n, y_n\}$ *to itself (possibly permuting them). Then there is an orientation-reversing homeomorphism of* $(\mathbb{R}^3, G)$ *that takes the collection of arcs* $A_1, A_2, \ldots, A_n$ *to itself (possibly permuting them).*

To understand how this theorem is relevant for us, first observe that if $G$ is an embedded cell complex and $h : (\mathbb{R}^3, G) \to (\mathbb{R}^3, G)$ is a homeomorphism, then

**Figure 4.17.** A cell attached at two points

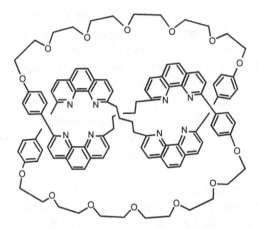

**Figure 4.18.** The molecular trefoil knot

*h* will take cells to cells. Furthermore, the vertices or edges that attach the cells to the rest of the cell complex will be sent to similar vertices or edges attaching the corresponding cells to the rest of the cell complex. Thus cells that are attached at one or more vertices, and cells that are attached along one or more edges, and cells that are attached along both vertices and edges, will be sent by *h* to cells that are attached in the same manner. For example, let *G* be an embedded cell complex and consider a cell *C* that is attached to *G* at two vertices, as is illustrated on the left in Figure 4.17. We shall refer to the two vertices where *C* is attached to *G* as the *attaching vertices* of *C*. Let *A* be the arc illustrated inside of *C* that has the two attaching vertices of *C* as its endpoints. Let *D* be some other cell that contains the arc *B* whose endpoints are the attaching vertices of *D*. Suppose there is an orientation-reversing homeomorphism of $(\mathbb{R}^3, G)$ that takes the cell *C* to the cell *D*. Then the homeomorphism will take the attaching vertices of *C* to the attaching vertices of *D*. Now by Theorem 4.3, there is an orientation-reversing homeomorphism of $(\mathbb{R}^3, G)$ that takes the arc *A* to the arc *B*.

We shall now see how to apply Theorem 4.3 to the molecular trefoil knot, which we illustrate in Figure 4.18. We can create a molecular cell complex *G* by replacing each isolated hexagon by a cell and each chain of three fused hexagons by a single cell. We prove by contradiction that our molecular cell

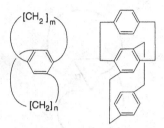

**Figure 4.19.** [$m$][$n$]paracyclophane and triple-layered cyclophane

complex is topologically chiral. Suppose that there is an orientation-reversing homeomorphism $h$ of ($\mathbb{R}^3$, $G$). Then $h$ takes each cell of the complex to a cell of the complex. The homeomorphism must take any attaching vertex of a cell to an attaching vertex of a cell. Now within each cell of the complex, choose an arc going between the two attaching vertices of that cell. By Theorem 4.3, there is an orientation-reversing homeomorphism $f$ of ($\mathbb{R}^3$, $G$) that takes the set of arcs in the cells of $G$ to itself. Now we create a trefoil knot $H$ by removing all of the cells and replacing them with these arcs. Then $H$ has the property that $f(H) = H$. However, this implies that a trefoil knot is topologically achiral. Because we know that the trefoil knot is topologically chiral, this gives us a contradiction. Thus the embedded cell complex of the molecular trefoil knot was, in fact, topologically chiral.

Observe that by considering molecular cell complexes, rather than molecular graphs, we end up being able to replace the impenetrable rings by edges (as our intuition tells us we should), and this replacement is now mathematically justified. Nonetheless, by using Kauffman's technique, in Chapter 2 we were able to prove the topological chirality of the graph of the molecular trefoil knot with an equal level of mathematical rigor. So, it may appear that the choice of using molecular cell complexes rather than molecular graphs appears to be more stylistic than substantive. However, we will now present some chemically chiral molecules whose molecular graphs are topologically achiral but whose molecular cell complexes are topologically chiral.

Consider the cell complexes of the molecules [$m$][$n$]paracyclophane (Nakazaki, Yamamoto, & Tanaka, 1971) and triple-layered cyclophane (Otsubo et al., 1971), which are illustrated on the left and the right, respectively, of Figure 4.19. The $m$ and the $n$ in the figure indicate that the $CH_2$ is repeated $m$ and $n$ times, respectively.

We shall prove that, for any $n$ and $m$, the cell complex of [$m$][$n$] paracyclophane is intrinsically chiral. Suppose there is some embedding of the cell complex of [$m$][$n$]paracyclophane that is topologically achiral. Let $G$ denote

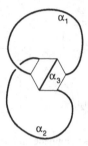

**Figure 4.20.** The cell complex of $[m][n]$paracyclophane together with arcs $\alpha_1$, $\alpha_2$, and $\alpha_3$

this embedded cell complex. Observe that $G$ has only one cell and this cell has two vertices that each have valence two and four vertices that each have valence three. We shall denote this cell by $C$, and the two vertices of valence two by $x$ and $y$. Let $\alpha_1$ and $\alpha_2$ denote the two arcs with endpoints on $C$ that contain the $CH_2$ chains. Let $\alpha_3$ denote a non–self-intersecting arc contained in $C$ with endpoints $x$ and $y$. Figure 4.20 illustrates the original embedding of the molecular cell complex of $[m][n]$paracyclophane, where the arcs $\alpha_1$, $\alpha_2$, and $\alpha_3$ have been darkened. By our assumption there is an orientation-reversing homeomorphism of $(\mathbb{R}^3, G)$ taking some embedding of the cell complex of $[m][n]$paracyclophane to itself. This homeomorphism takes $C$ to itself and the pair of points $\{x, y\}$ to itself, possibly interchanging them. Thus by Theorem 4.3, there is an orientation-reversing homeomorphism $h$ of $(\mathbb{R}^3, G)$ taking the cell complex to itself that takes the arc $\alpha_3$ to itself. Furthermore, $h$ takes the pair of arcs $\{\alpha_1, \alpha_2\}$ to itself, possibly interchanging them. Let $K$ denote the boundary of the cell $C$; then $h(K) = K$. Now let $M$ denote the embedded three-rung Möbius ladder consisting of $K$ as the loop, together with $\alpha_1$, $\alpha_2$, and $\alpha_3$ as the rungs. Then $h$ is an orientation-reversing homeomorphism of $(\mathbb{R}^3, M)$ that takes $K$ to itself. This contradicts Corollary 3.4. Therefore the cell complex of $[m][n]$paracyclophane was intrinsically chiral. The argument for triple-layered cyclophane is sufficiently similar that we leave it to the reader.

In contrast, with the molecular cell complexes, the molecular graphs of $[m][n]$paracyclophane and triple-layered cyclophane can be deformed into a plane. We illustrate planar embeddings of these molecular graphs in Figure 4.21. Any graph that can be deformed into a plane is topologically achiral, so the molecular bond graphs of both $[m][n]$paracyclophane and triple-layered cyclophane are topologically achiral. We just saw that these cell complexes are intrinsically chiral, in particular, the molecular cell complexes of both of these molecules are topologically chiral. Chemically, both $[m][n]$paracyclophane and triple-layered cyclophane are chiral. So, the chirality of these molecules is better represented by their molecular cell complexes than by their molecular graphs.

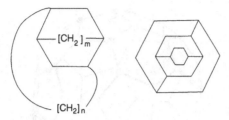

**Figure 4.21.** Planar presentations of the graphs of $[m][n]$paracyclophane and triple-layered cyclophane

We can use the same method to show that the molecular cell complex of triple-layered naphthalenophane (illustrated in Figure 3.31 of Chapter 3) is intrinsically chiral. In fact, we can create cell complexes $C(p, q, r)$ analogous to the graphs $G(p, q, r)$ of Chapter 3 (Figure 3.32) by replacing each hexagon by a cell. Note that the cell complex of triple-layered cyclophane is just $C(1, 1, 1)$. We can use the above method to prove that for any $p$, $q$, and $r$, the cell complex $C(p, q, r)$ is intrinsically chiral. This is in contrast with the result we obtained in Chapter 3 that $G(p, q, r)$ is intrinsically chiral if and only if all of $p$, $q$, and $r$ are even. Note that the achiral embeddings that we obtained in Chapter 3, where at least one of $p$, $q$, or $r$ is odd, each contained at least one benzene ring that has a piece of the graph running through it. On a chemical level this cannot happen; thus again the cell complex $C(p, q, r)$ more accurately represents the structure of these hypothetical molecules than does the graph $G(p, q, r)$.

We can see from the examples given above that a molecular graph may be topologically achiral, whereas the corresponding molecular cell complex is topologically chiral and the molecule itself is chiral. Furthermore, if the cell complex of a molecule is topologically achiral, then the graph of that molecule is also topologically achiral. All of the topological information that a molecular graph contains about the molecule is also contained in the molecular cell complex, but in addition the molecular cell complex contains the information that certain types of molecular deformations cannot occur because of impenetrability. In this way we see that molecular cell complexes can be an important tool in understanding when the topology of a molecule causes it to be chiral.

Another use for molecular cell complexes is to represent molecules whose structures resemble a surface. For example, we can consider the family of fullerenes. These molecules are made up of many small rings of carbon atoms that are fused together along all of their boundaries so as to create the structure of a surface like that created by chicken wire. Figure 4.22 represents the molecular cell complex of the fullerene $C_{60}$, which contains 60 carbon atoms. This molecule has the structure of a truncated icosahedron, which resembles the pattern seen on a soccer ball. Fullerenes are a new form of carbon molecules,

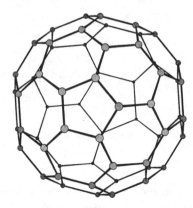

**Figure 4.22.** The molecular graph of the fullerene $C_{60}$

which have been useful in superconductors and in the production of tiny diamonds. Chemists are quite interested in studying fullerenes, both because of their potential applications and because of the aesthetics of their incredible symmetry. Fullerenes in the shape of a sphere have been synthesized (Krätschmer et al., 1990) and also have been discovered to occur naturally in a rock called shungite, which is found in Russia (Buseck, Tsipursky, & Hettich, 1992). In addition, fullerenes in the form of a torus have recently been observed (Liu et al., 1997). In all of their forms, fullerenes behave as surfaces because the small rings making up these structures are each impenetrable. Thus, representing fullerenes as embedded graphs will not convey the topology of these structures as well as representing them as cell complexes, where each ring is represented by a cell.

## Exercises

1. Is the figure eight knot a topological rubber glove in $\mathbb{R}^3$? Explain.
2. Explain why the knot $8_{17}$ is a topological rubber glove in $\mathbb{R}^3$ but not in $S^3$?
3. Prove that there is no orientation preserving homeomorphism of $S^3$ which takes the knot $8_{17}$ to itself, reversing the orientation on the knot.
4. Explain why the DNA figure eight knot is a topological rubber glove? Is it chemically achiral?
5. Suppose $h$ is a finite order homeomorphism of $S^3$ which pointwise fixes a sphere $P$ and takes the structure in Figure 4.12 to itself. Explain why this implies that the naphthalene at the top of the figure has to be pointwise fixed by $h$.
6. Explain why $[m][n]$paracyclophane is intrinsically chiral as a cell complex, but not topologically achiral as a graph?
7. Let $h : \mathbb{R}^3 \rightarrow \mathbb{R}^3$ and $f : \mathbb{R}^3 \rightarrow \mathbb{R}^3$. Suppose that the order of $h$ is $n$. Could the order of $f^{-1} \circ h \circ f$ be less than $n$? Explain.
8. Prove that the molecular connected sum, made of two right-handed molecular trefoil knots, is topologically chiral.

9. Suppose that $h : S^1 \to S^1$ is a finite-order homeomorphism that reverses the orientation of $S^1$. Prove that $h$ has an order of two.
10. Prove that a knot is negative achiral in $S^3$ if and only if it is negative achiral in $\mathbb{R}^3$.
11. Create an intrinsic topological rubber glove that contains a complete graph on six vertices, $K_6$. Prove that your example works.
12. Suppose that the molecule illustrated in Figure 4.12 had a benzene at the top rather than a naphthalene. Would it still be a topological rubber glove? Explain.

13. Show that the following embedded graph is a topological rubber glove in $S^3$.
14. If the bases in the DNA figure eight knot occurred in a palindromic sequence, would it be a topological rubber glove in $\mathbb{R}^3$? What about in $S^3$? Explain.
15. Prove that the cell complex of triple-layered cyclophane (see Figure 4.19) is intrinsically chiral.
16. Prove that the cell complex $C(p, q, r)$ is intrinsically chiral for every $p$, $q$, and $r$.
17. Suppose that $K$ is a knot that is both positive achiral and negative achiral. Prove that there exists an orientation-preserving homeomorphism $h : (S^3, K) \to (S^3, K)$ that reverses the orientation of $K$.
18. Find a topologically chiral knot $K$ that has the property that there is an orientation-preserving homeomorphism $h : (S^3, K) \to (S^3, K)$ such that $h$ reverses the orientation of $K$.
19. Find a composite knot that is positive achiral but not negative achiral. Find a composite knot that is negative achiral but not positive achiral.
20. Find a graph, different from the one illustrated in Figure 4.14, that has a rigidly achiral embedding in $S^3$ but no rigidly achiral embedding in $\mathbb{R}^3$.
21. Describe a finite-order orientation-reversing homeomorphism of the graph of L-alanine, illustrated below.

$$CH_3$$
$$|$$
$$H - \overset{\displaystyle C}{\underset{\displaystyle |}{\phantom{C}}} {}^{\prime\prime\prime\prime\prime\prime}CO_2H$$
$$NH_2$$

22. Prove Theorem 4.3.
23. Suppose $h : S^2 \to S^2$ has finite order. What are the possibilities for the fixed point set of $h$ if $h$ is orientation reversing? What are the possibilities for the fixed point set of $h$ is if $h$ is orientation preserving? Give an example of an $h$ for each type of fixed point set that you list.
24. Repeat Exercise 24, for $h : \mathbb{R}^2 \to \mathbb{R}^2$, $h : S^1 \to S^1$, and $h : \mathbb{R}^1 \to \mathbb{R}^1$.
25. Is there a finite order orientation reversing homeomorphism of $\mathbb{R}^3$ which pointwise fixes the fullerene illustrated in Figure 4.22? Explain.

# 5

# Intrinsic Topological Properties of Graphs

We have discussed the concept of intrinsic chirality in Chapters 3 and 4, and we have seen various examples of molecular graphs and cell complexes that are intrinsically chiral. Recall that we call such graphs *intrinsically chiral* because their chirality is an intrinsic property of the graphs rather than a result of how the graphs are embedded in $\mathbb{R}^3$ or $S^3$. Although there are many other intrinsic properties of abstract graphs, we are interested in those properties that tell us something about the topology of all possible embeddings of the graphs in $\mathbb{R}^3$ or $S^3$. One class of graphs that has several intrinsic topological properties is the class of complete graphs. We saw before that a *complete graph* on $n$ vertices, denoted by $K_n$, is defined to be $n$ vertices together with edges connecting every pair of vertices. For example, in Chapter 1 we saw the graph of the Simmons–Paquette molecule. This graph (illustrated in Figure 5.1) has the form of a $K_5$ together with some additional vertices of valence two.

Liang and Mislow (1994a) have listed many other molecules, including a protein molecule, whose graphs also contain a $K_5$. Thus far, no molecule has been synthesized whose molecular graph is $K_n$ for $n > 5$. Sritana-Anant, Seiders, and Siegel (1998) have proposed the graph $K_6$ as well as other complex graphs as targets for future molecular design and synthesis. Thus we expect that in the future we will see a molecular $K_6$. Seeman (1999) has used synthetic DNA to create many complex molecular knots, links, and graphs. Similar techniques may also be used in the future to create various complete graphs out of synthetic DNA. Even using synthetic DNA, as $n$ gets larger the difficulty of synthesizing a molecular $K_n$ increases. Currently we do not know what the largest $n$ is for which a molecular $K_n$ could ever be synthesized.

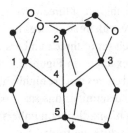

**Figure 5.1.** The Simmons–Paquette molecule

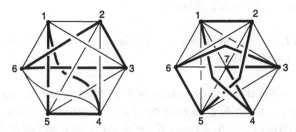

**Figure 5.2.** Every embedding of $K_6$ contains a nontrivial link, and every embedding of $K_7$ contains a nontrivial knot

## Results of Conway and Gordon

In Chapter 1, we saw that $K_5$ has the intrinsic property that it cannot be embedded in a plane. Because the complete graphs $K_6$ and $K_7$ contain $K_5$, they are also are nonplanar. However, in addition, they have other intrinsic properties that are topologically interesting. In particular, Conway and Gordon (1983) proved the surprising result that every embedding of $K_6$ in $\mathbb{R}^3$ contains a nontrivial link and every embedding of $K_7$ in $\mathbb{R}^3$ contains a nontrivial knot. We illustrate embeddings of $K_6$ and $K_7$ in Figure 5.2, with a link and a knot highlighted in $K_6$ and $K_7$, respectively. A graph that has the property that every embedding of it contains a nontrivial link is said to be *intrinsically linked*, and a graph that has the property that every embedding of it contains a nontrivial knot is said to be *intrinsically knotted*. Intrinsic knotting and linking are examples of intrinsic properties of a graph that have an effect on the extrinsic topology of any embedding of the graph in $\mathbb{R}^3$.

Before we discuss the proofs of Conway and Gordon, we shall begin by observing that any set of three vertices in $K_6$ determines a simple closed curve consisting of these three vertices together with the edges between them.

Because $K_6$ has six vertices, this set of three vertices also determines a complementary simple closed curve with three vertices in $K_6$. We number the vertices, one through six, and let each set of three numbers represent a pair of disjoint simple closed curves. For example, in Figure 5.2 the set of vertices $\{1, 4, 5\}$ determines the pair of linked curves that are highlighted. Note, however, that the set of vertices $\{2, 3, 6\}$ also determines this same pair of linked curves. There are twenty triples made up of six distinct numbers. However, as each pair of curves is determined by two different triples, we see that $K_6$ has a total of ten pairs of disjoint curves. We will sketch the proofs of Conway and Gordon's results rather than presenting complete proofs in order to convey their ideas without including too many technical details.

**Theorem 5.1.** *Every embedding of* $K_6$ *in* $\mathbb{R}^3$ *contains a nontrivial link (Conway & Gordon, 1983).*

*Sketch of Proof.* Let $K_6$ be embedded in $\mathbb{R}^3$. For each pair of disjoint loops $A$ and $B$ in $K_6$, let $\omega(A, B)$ denote the linking number of $A$ and $B$ considered mod 2. That is, $\omega(A, B) = 1$ if $\mathrm{Lk}(A, B)$ is odd, and $\omega(A, B) = 0$ if $\mathrm{Lk}(A, B)$ is even. Here the choice of orientation for $A$ and $B$ is irrelevant, as it will not affect the parity of $\mathrm{Lk}(A, B)$ (the reader should prove this). Let $\omega = \sum_{A,B} \omega(A, B) \bmod 2$. That is, $\omega$ is the mod 2 sum of the linking numbers of all pairs of disjoint loops contained in $K_6$. So, for a given embedding, either $\omega = 0$ or $\omega = 1$.

Every embedding of $K_6$ in $\mathbb{R}^3$ can be obtained from a given embedding by an ambient isotopy of the graph together with changing crossings from overcrossings to undercrossings or from undercrossings to overcrossings. Because the linking number is a topological invariant, ambient isotopy does not change the value of $\omega$. Conway and Gordon showed that if we change any crossing in an embedding of $K_6$, then it will add one to an even number of the mod 2 linking numbers $\omega(A, B)$, and thus have no effect on the mod 2 sum $\omega$. Therefore, $\omega$ has the same value, no matter how $K_6$ is embedded in $\mathbb{R}^3$. Computing $\omega$ for the particular embedding of $K_6$ that is illustrated in Figure 5.1 yields $\omega = 1$. If for some embedding, every link in $K_6$ were trivial, then that embedding would have $\omega = 0$, as the linking number of a trivial link of two components is zero. Because $\omega = 1$ for every embedding, it follows that every embedding of $K_6$ in $\mathbb{R}^3$ contains a nontrivial link.                                       $\square$

The idea of the proof that every embedding of $K_7$ in $\mathbb{R}^3$ contains a nontrivial knot is based on a similar idea as the above proof; however, instead of

**Figure 5.3.** The two types of pass moves

using linking numbers, Conway and Gordon make use of the *Arf invariant* of knots. There are several different ways to define the Arf invariant, which are all equivalent. We shall present the most elementary definition. Given an oriented knot, the act of passing one pair of strings with opposite orientations through another such pair (as illustrated in Figure 5.3) to obtain a new knot is called a *pass move*. As most knots do not have parallel strands at each crossing, a knot may first have to be deformed in order to perform such a pass move. Two oriented knots are said to be *pass equivalent* if we can get from one knot to the other by means of a sequence of pass moves together with deformations. It can be proven that every knot is either pass equivalent to the unknot or to a trefoil knot, and the unknot is not pass equivalent to a trefoil knot (see Kauffman, 1983). The Arf invariant of a knot $K$, denoted by $a(K)$, is defined to be equal to zero if $K$ is pass equivalent to the unknot, and it is defined to be equal to one otherwise. By definition, two knots that are ambient isotopic will have the same Arf invariant. Thus the Arf invariant is a topological invariant. Observe that as with the P-polynomial, the particular choice of orientation on a knot $K$ does not affect the value of $a(K)$, as changing the orientation on the knot changes all of the arrows involved in each pass move. Also, it is not hard to prove that a knot and its mirror image always have the same Arf invariant. This means that the Arf invariant will not be useful in proving the topological chirality of knots.

In Chapter 2 we saw that the P-polynomial satisfies the equation $lP(L_+) + l^{-1}P(L_-) + mP(L_0) = 0$, where the oriented links $L_+, L_-$, and $L_0$ are identical except at a single crossing where $L_+$ has a positive crossing, $L_-$ has a negative crossing, and $L_0$ has a null crossing. The Arf invariant satisfies a similar type of equation. In particular, if $K$ is an oriented knot then changing a crossing will create a (possibly different) knot, and smoothing a crossing will create a two-component link. So, we can let $K_+$, $K_-$, and $L = L_1 \cup L_2$ be identical except at the crossing that is illustrated in Figure 5.4. Kauffman (1983) showed that the Arf invariant as we defined it satisfies the equation $a(K_+) = a(K_-) + \text{Lk}(L_1, L_2)$. This equation is important in Conway and Gordon's proof that every embedding of $K_7$ in $\mathbb{R}^3$ contains a nontrivial knot.

Rather than looking at simple closed curves in $K_7$ of arbitrary length, Conway and Gordon focus on only those simple closed curves that go through every

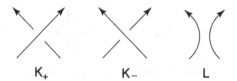

$K_+$        $K_-$        $L$

**Figure 5.4.** $K_+$, $K_-$, and $L = L_1 \cup L_2$ are identical except at the above crossing

vertex of the graph. Such a curve is called a *Hamiltonian cycle* of the graph. Because there is an edge between every pair of vertices in $K_7$, any ordered list of all seven vertices will determine a Hamiltonian cycle of $K_7$. For example, we can specify the knot that is highlighted in Figure 5.2 by the ordered list $(1, 2, 5, 6, 3, 7, 4)$. However, there are fourteen ways to list this same simple closed curve, which we can obtain by cyclically permuting the numbers in the list and by reversing them. Thus there are a total of $7!/14 = 360$ Hamiltonian cycles in $K_7$. Below we very briefly sketch the proof of Conway and Gordon's (1983) result about $K_7$.

**Theorem 5.2.** *Every embedding of* $K_7$ *in* $\mathbb{R}^3$ *has a Hamiltonian cycle that contains a nontrivial knot.*

*Sketch of Proof.* Let $K_7$ be embedded in $\mathbb{R}^3$. Let $C$ denote the set of all Hamiltonian cycles in $K_7$. Let $a = \sum_{K \in C} a(K) \bmod 2$. That is, $a$ is the mod 2 sum of the Arf invariants of all of the Hamiltonian cycles in $K_7$. Every embedding of $K_7$ in $\mathbb{R}^3$ can be obtained from a given embedding by ambient isotopy together with crossing changes. Using the equation $a(K_+) = a(K_-) + \mathrm{Lk}(L_1, L_2)$, Conway and Gordon proved that if any crossing of $K_7$ is changed, it will not change the value of $a$. If there were an embedding of $K_7$ that did not contain any knots, then the Arf invariant of every Hamiltonian cycle in that embedding would be zero. Computing $a$ for the particular embedding of $K_7$ illustrated in Figure 5.1 yields $a = 1$. Thus every embedding of $K_7$ has $a = 1$, and hence it must contain at least one nontrivial knot.      □

## Minor Minimal Intrinsically Linked Graphs

It is not hard to see that if $G$ is a graph that contains $K_6$ as a subgraph, then $G$ will also be intrinsically linked. Similarly, if $G$ is a graph that contains $K_7$, then $G$ will be intrinsically knotted. Thus, for every $n > 6$, the complete graph $K_n$ is both intrinsically linked and intrinsically knotted. It would be nice to have a list of the simplest possible graphs that are intrinsically linked or intrinsically knotted. We formalize the notion of *simplest* by using the concept of minors.

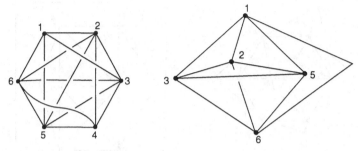

**Figure 5.5.** Unlinked embeddings of $K_6$ with an edge deleted and $K_6$ with an edge contracted

If $G$ is a graph, we say that $H$ is a *minor* of $G$ if $H$ can be obtained from $G$ by deleting or contracting a finite number of edges of $G$. For example, $K_6$ is a minor of $K_7$ because if we start with $K_7$ and delete all of the edges containing vertex 7 except for one, and contract the remaining edge containing vertex 7, we will obtain $K_6$. It was shown by Motwani, Raghunathan, & Saran (1988) that if a graph $H$ is a minor of a graph $G$, and $H$ is intrinsically linked or intrinsically knotted, then $G$ is intrinsically linked or intrinsically knotted as well. Observe that intrinsic chirality does not have the analogous property. For example, the Möbius ladder $M_5$ is intrinsically chiral, and $M_5$ is a minor of $M_6$. However, $M_6$ is not intrinsically chiral.

We say that a graph $G$ is *minor minimal* with respect to a certain property if $G$ has that particular property, but no minor of $G$ has the property. We let $G - e$ denote $G$ with the edge $e$ deleted, and we let $G/e$ denote $G$ with the edge $e$ contracted. In order to determine whether $K_6$ is minor minimal with respect to being intrinsically linked, we delete a single edge $e$ of $K_6$ to see if there exists an embedding of $K_6 - e$ in $\mathbb{R}^3$ that contains no link. We also contract a single edge $e$ of $K_6$ to see if there is an embedding of the contracted graph $K_6/e$ that contains no link. For any pair of edges there is an automorphism of $K_6$ that interchanges the two edges. Hence it makes no difference which edge we delete or contract. On the left of Figure 5.5, we have deleted the edge $\overline{14}$ and embedded the rest of the graph as we did in Figure 5.3. The highlighted link in Figure 5.1 was the only nontrivial link in this embedding, so the graph on the left of Figure 5.5 contains no nontrivial link. On the right in Figure 5.5 we have created an embedding of $K_6$ with the edge $\overline{34}$ contracted. By contracting this edge we end up with two edges connecting the vertex 3 to every other vertex. So that we will still have a graph, we omit one edge from each pair of such edges. The reader can see by inspection that this graph is a $K_6$ which contains no nontrivial link. Now any minor of $K_6$ is actually a minor of one of the two

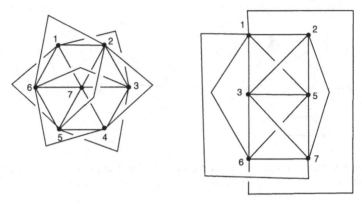

**Figure 5.6.** Unknotted embeddings of $K_7$ with an edge deleted and $K_7$ with an edge contracted

graphs in Figure 5.5, and hence it has an embedding with no nontrivial link. Thus $K_6$ is minor minimal with respect to being intrinsically linked.

Similarly, the left side of Figure 5.6 shows an embedding of $K_7$ with the edge $\overline{14}$ deleted and the right side of Figure 5.6 shows an embedding of $K_7$ with the edge $\overline{34}$ contracted. The reader should check that neither of these embedded graphs contains a knot. Now any minor of $K_7$ is a minor of one of these two graphs. Hence it follows that any minor of $K_7$ has an unknotted embedding. Thus $K_7$ is minor minimal with respect to being intrinsically knotted.

Sachs (1983) used a method similar to that of Conway and Gordon to prove that $K_6$ as well as the six graphs in Figure 5.7 are all intrinsically linked, and that none of the minors of these graphs are intrinsically linked. This set of graphs is called the *Peterson family* of graphs.

Sachs conjectured that the graphs in the Peterson family are the only graphs that are minor minimal with respect to being intrinsically linked. Sachs' conjecture was proved in 1995 by Robertson, Seymour, and Thomas (Robertson et al., 1995 ). In fact, Robertson et al. (1995) proved the following stronger result, which proves Sach's conjecture as a consequence.

**Theorem 5.3.** *Let* G *be a graph. Then the following are equivalent:* (1) G *has an embedding in* $\mathbb{R}^3$ *that contains no nontrivial link;* (2) G *has an embedding in* $\mathbb{R}^3$ *such that every simple closed curve contained in* G *bounds a disk in the complement of* G; *and* (3) G *has no minor in the Peterson family.*

Theorem 5.3 is quite a surprising result, because every embedded graph that contains no link will not necessarily satisfy condition 2. For example,

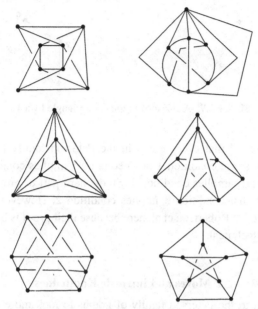

**Figure 5.7.** Other graphs that are minor minimal with respect to being intrinsically linked

**Figure 5.8.** Kinoshita's theta curve

consider the embedded graph that is shown in Figure 5.8. This embedded graph is known as *Kinoshita's theta curve* (see Kinoshita, 1972). It contains no knots or links, yet none of the three simple closed curves in the graph bounds a disk in the complement of the graph. Of course this graph can be re-embedded as a planar circle together with a diameter, so that it looks like the Greek letter $\theta$, in which case all three curves bound disks whose interiors are disjoint from the graph. This example highlights the fact that Theorem 5.3 is *not* saying that any embedded graph that has no nontrivial link has the property that every cycle bounds a disk disjoint from the graph. Rather, Theorem 5.3 says that if a graph has an embedding that contains no nontrivial link, then it has a (possibly different) embedding such that every cycle bounds a disk disjoint from the graph.

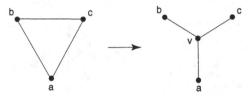

**Figure 5.9.** A $\Delta Y$ move replaces a triangle by a $Y$

As Sachs proved that every graph in the Peterson family is intrinsically linked, it follows that condition 1 of Theorem 5.3 implies condition 3. Also, condition 2 clearly implies condition 1. So the hard part of proving Theorem 5.3 is showing that condition 3 implies condition 2. However, we will not present the proof of Robertson et al. here because their proof is both extremely long and quite technical.

### $\Delta Y$ Moves and Intrinsic Knottedness

Now we return to the Peterson family of graphs to look more closely at the relationship between these graphs, and we discuss some results about intrinsic knottedness. We begin with a definition for abstract graphs.

**Definition.** Suppose that a graph $G$ contains a simple closed curve with precisely three vertices, say $a$, $b$, and $c$. Let $G'$ denote the graph that is identical to $G$ except that the edges $\overline{ab}$, $\overline{ac}$, and $\overline{bc}$ have been replaced by a new vertex $v$ of valence 3 together with three new edges $\overline{av}$, $\overline{bv}$, and $\overline{cv}$. Then we say that $G'$ has been obtained from $G$ by a $\Delta Y$ move and that $G$ has been obtained from $G'$ by a $Y\Delta$ move.

Figure 5.9 illustrates a triangle that is replaced by a $Y$ when we do a $\Delta Y$ move.

The graphs illustrated in Figure 5.7 are all the graphs that can be obtained from $K_6$ by doing a finite number of $\Delta Y$ moves and $Y\Delta$ moves (the reader should check this). It is natural to ask whether the set of graphs that are minor minimal with respect to being intrinsically knotted are precisely those that can be obtained from $K_7$ by $\Delta Y$ moves and $Y\Delta$ moves. Motwani et al. (1988) prove that $\Delta Y$ moves preserve the properties of being intrinsically linked and intrinsically knotted. More specifically, they prove the following Lemma.

**Lemma 5.4.** *Let G be a graph and suppose that G' is obtained from G by a $\Delta Y$ move. If G is intrinsically linked then G' is intrinsically linked, and if G is intrinsically knotted then G' is intrinsically knotted.*

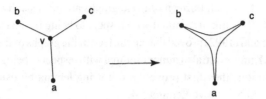

**Figure 5.10.** To embed $G$ we replace the $Y$ in $G'$ by a nearby triangle

*Proof.* We will prove this for intrinsically knotted graphs; the proof for intrinsically linked graphs is essentially the same. Suppose that $G$ is intrinsically knotted but $G'$ is not intrinsically knotted, and $G'$ is obtained from $G$ by replacing the triangle that has edges $\overline{ab}$, $\overline{ac}$, and $\overline{bc}$ by the $Y$ that has edges $\overline{av}$, $\overline{bv}$, and $\overline{cv}$. By hypothesis, $G'$ has an embedding in $\mathbb{R}^3$ such that no simple closed curve in $G'$ is knotted. Start with this embedding of $G'$ and choose a small tube in the shape of a $Y$, which contains the edges $\overline{av}$, $\overline{bv}$, and $\overline{cv}$ and does not intersect any other part of the graph. We create an embedding of $G$ from the embedding of $G'$ by removing the edges $\overline{av}$, $\overline{bv}$, and $\overline{cv}$ and replacing them by a triangle with edges $\overline{ab}$, $\overline{ac}$, and $\overline{bc}$, which lies within the $Y$-shaped tube and looks like the triangle illustrated in Figure 5.10.

Now, because we know that $G$ is intrinsically knotted, there must be a knotted simple closed curve $K$ that is contained in this embedding of $G$. If $K$ does not contain any of the edges $\overline{ab}$, $\overline{ac}$, or $\overline{bc}$, then the knot $K$ is also contained in the embedded graph $G'$, which is contrary to hypothesis. Thus $K$ must contain at least one of $\overline{ab}$, $\overline{ac}$, and $\overline{bc}$. Suppose that $K$ contains just one of these three edges. Without loss of generality, say that $K$ contains $\overline{ab}$. Now define $K'$ to be the simple closed curve in $G'$ that is identical to $K$ except that the edge $\overline{ab}$ in $G$ has been replaced by the two edges $\overline{av}$ and $\overline{vb}$ in $G'$. By the construction of our embedding of $G$, the curve $K'$ will be ambient isotopic to $K$ in $\mathbb{R}^3$. Hence $K'$ is a nontrivial knot contained in $G'$, again contrary to hypothesis. Next suppose that $K$ contains two edges of the triangle $\triangle abc$. Without loss of generality, $K$ contains $\overline{ab}$ and $\overline{bc}$. In this case, we define a new simple closed curve $K''$, which is identical to $K$ except the two edges $\overline{ab}$ and $\overline{bc}$ have been replaced by the two edges $\overline{av}$ and $\overline{vc}$. However, again $K''$ is ambient isotopic to $K$ in $\mathbb{R}^3$, and thus $K''$ is knotted, contrary to hypothesis. If $K$ were to contain all three edges $\overline{ab}$, $\overline{ac}$, and $\overline{bc}$, then $K$ would not be knotted, because by construction this triangle was not knotted. Therefore, the graph $G'$ must, in fact, be intrinsically knotted. $\square$

If we start with $K_7$ and repeatedly apply $\Delta Y$ moves, we obtain thirteen additional graphs. We do not illustrate these graphs here [see Kohara & Suzuki (1992) for an illustration of the thirteen graphs]. It follows from Lemma 5.4

that in addition to $K_7$, all thirteen of these graphs are intrinsically knotted. We saw above that $K_7$ is minor minimal with respect to being intrinsically knotted. Kohara and Suzuki (1992) proved that all thirteen of the graphs obtained from $K_7$ by applying $\Delta Y$ moves are also minor minimal with respect to being intrinsically knotted. To do this, they first prove the following lemma by using a method similar to that used to prove Lemma 5.4.

**Lemma 5.5.** *Let* G *be a graph that is minor minimal with respect to being intrinsically knotted, and let* G′ *be obtained from* G *by a* $\Delta Y$ *move at the triangle* $\triangle abc$. *Suppose that for each edge* e *of* G *that is not one of the edges* $\overline{ab}$, $\overline{ac}$, *or* $\overline{bc}$, *we have both of the following: (1) there is an embedding of* G − e *that contains no knot and is such that the triangle* $\triangle abc$ *bounds a disk whose interior is disjoint from the embedding of* G − e; *(2) there is an embedding of* G/e *that contains no knot and is such that the triangle* $\triangle abc$ *bounds a disk whose interior is disjoint from the embedding of* G/e. *Then* G′ *is minor minimal with respect to being intrinsically knotted.*

*Proof.* First note that we know by Lemma 5.4 that $G'$ is intrinsically knotted. We need to show that if $H$ is a minor of $G'$ then there is an embedding of $H$ that contains no knot. We will show that for any edge $e$ of $G'$, there are embeddings of $G' - e$ and $G'/e$ that contain no knots. Any minor of $G'$ is either $G' - e$, $G'/e$, or is a minor of $G' - e$ or $G'/e$ for some edge $e$. Thus if we show that for any edge $e$, there are embeddings of $G' - e$ and $G'/e$ that contain no knots, then every minor of $G'$ will have an embedding that contains no knots.

Let $e$ be an edge of $G'$. First suppose that $e$ is one of the edges $\overline{av}$, $\overline{bv}$, or $\overline{cv}$. Without loss of generality, $e = \overline{av}$. Let $H$ denote the graph $G$ with the two edges $\overline{ab}$ and $\overline{ac}$ removed. The graphs $G' - e$ and $H$ are homeomorphic as sets, though as graphs the arc in $G' - e$ going from $b$ to $c$ consists of the edges $\overline{bv}$ and $\overline{vc}$ whereas the arc in $H$ from $b$ to $c$ is a single edge. Now $H$ is a minor of $G$ and so has an embedding that is not knotted. We create an embedding of $G' - e$ by putting the vertex $v$ anywhere in the interior of the edge $\overline{bc}$ in the embedding of $H$. This embedding of $G' - e$ contains no knot. Next, observe that the graph $G'/e$ is identical to the graph $G - \overline{bc}$, which is a minor of $G$ and so has an unknotted embedding.

Now we suppose that $e$ is not one of the edges $\overline{av}$, $\overline{bv}$, or $\overline{cv}$. Then $e$ is also an edge of $G$ that is not one of the edges $\overline{ab}$, $\overline{ac}$, or $\overline{bc}$. By hypothesis, there is an embedding of $G - e$ that contains no knot and such that the triangle $\triangle abc$ bounds a disk $D$ whose interior is disjoint from $G - e$. We create an embedding of $G' - e$ by starting with this embedding of $G - e$; then we remove the edges $\overline{ab}, \overline{ac}$, and $\overline{bc}$ and within the disk $D$ we add the vertex $v$ and the edges $\overline{av}, \overline{bv}$, and $\overline{cv}$. Suppose that this embedding of $G' - e$ contains a knot $K'$. Because there is no knot in $G - e$, the knot $K'$ must contain at least one of the edges $\overline{av}, \overline{bv}$, or $\overline{cv}$. However, because $K'$ is a simple closed curve, $K'$ cannot contain only one of these edges. Also, because $K'$ is knotted, $K'$ cannot contain all three of these edges. Hence $K'$ contains precisely two of these edges, say $\overline{av}$ and $\overline{vb}$. But now there is a simple close curve $K$ in $G - e$ that is identical to $K'$ except that the two edges $\overline{av}$ and $\overline{vb}$ have been replaced by the single edge $\overline{ab}$. Furthermore, by our construction of the embedding of $G' - e$, we must have that $K$ is ambient isotopic to $K'$ in $\mathbb{R}^3$, and hence $K$ is knotted. As this is contrary to hypothesis, this embedding of $G' - e$ could contain no nontrivial knot. The argument that there is an embedding of $G'/e$ that contains no knot is identical.

Therefore, every minor of $G'$ has an embedding that contains no knot.  □

Note that an analogous lemma can be proven for graphs that are minor minimal with respect to the property of being intrinsically linked (Kohara & Suzuki, 1992).

**Theorem 5.6.** *The thirteen graphs that can be obtained from* $K_7$ *by* ΔY *moves are all minor minimal with respect to the property of being intrinsically knotted.*

We do not prove Theorem 5.6, because the proof involves too many cases. We observe, however, that four of the thirteen graphs can be obtained from one of the other nine graphs by a single ΔY move. Thus for each of these nine graphs $G$, we find unknotted embeddings for all of the minors $G - e$ and $G/e$ that each have the property that the triangle on which we will do a ΔY move bounds a disk in the complement of the minor; then Lemma 5.3 implies that the remaining four graphs are also minor minimal with respect to being intrinsically knotted.

At the present time, it is not known whether $K_7$ together with the thirteen graphs that can be obtained from it by doing ΔY moves are the only graphs that are minor minimal with respect to being intrinsically knotted. An important consequence of Theorem 5.3 is that any graph that is intrinsically knotted is also intrinsically linked. This follows because condition 2 of Theorem 5.3 implies that $G$ has an embedding that does not contain a knot. Thus if a graph has an embedding that does not contain a nontrivial link, then it has an embedding that does not contain a nontrivial knot. Hence every graph that is minor minimal with respect to being intrinsically knotted contains a Peterson graph as a minor. However, none of the Peterson graphs themselves are intrinsically knotted (the reader should check this). It was conjectured by Motwani et al. (1988) and by Kohara and Suzuki (1992) that the graph $K_{3,3,1,1}$ is minor minimal with respect to being intrinsically knotted. This graph is not one of the thirteen graphs that can

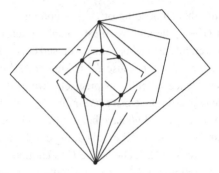

**Figure 5.11.** An embedding of $K_{3,3,1,1}$ that contains exactly two trefoils as Hamiltonian cycles

be obtained from $K_7$ by doing $\Delta Y$ moves. Figure 5.11 illustrates an embedding of $K_{3,3,1,1}$ that contains exactly two trefoils as Hamiltonian cycles. Also note that we can obtain $K_{3,3,1,1}$ from the second graph in Figure 5.7 (which is $K_{3,3,1}$) by adding a vertex at the bottom and edges connecting this vertex to all of the other vertices. Kohara and Suzuki (1992) observed that there are twenty-five other graphs that can be obtained from $K_{3,3,1,1}$ by doing $\Delta Y$ moves. Thus if $K_{3,3,1,1}$ is intrinsically knotted, then these other twenty-five graphs would also be intrinsically knotted by Lemma 5.2.

### Intrinsic Chirality of Complete Graphs

Now we return to the concept of intrinsic chirality and make use of the work of Conway and Gordon to prove that any complete graph on $4n + 3$ vertices is intrinsically chiral. That is, we shall show that the graphs $K_7$, $K_{11}$, $K_{15}$, ... are all intrinsically chiral.

**Definition.** Let $G$ be a graph and let $\Phi$ be an automorphism of $G$. We say that $\Phi$ is *induced* or *realized* by a homeomorphism $h : (\mathbb{R}^3, G) \to (\mathbb{R}^3, G)$ if the action of $h$ on the vertices of $G$ is identical to that of $\Phi$.

Note that if $h : (\mathbb{R}^3, G) \to (\mathbb{R}^3, G)$, then by our definition (see Chapter 2), $h$ takes vertices to vertices and edges to edges, so $h$ necessarily induces some automorphism on the vertices of $G$.

**Lemma 5.7.** *Let* $K_7$ *be embedded in* $\mathbb{R}^3$ *and let* $f : (\mathbb{R}^3, K_7) \to (\mathbb{R}^3, K_7)$ *be a homeomorphism. If* $f$ *induces an order-two automorphism of the vertices of* $K_7$, *then* $f$ *fixes precisely one vertex of* $K_7$ *(Flapan & Weaver, 1992).*

*Proof.* As in the proof of Theorem 5.2, for a given embedding of $K_7$, we let $C$ denote the set of all Hamiltonian cycles in $K_7$. For each $K \in C$, we let $a(K)$ denote the Arf invariant of $K$, and we let $a = \sum_{K \in C} a(K) \bmod 2$. That is, $a$ is the mod 2 sum of the Arf invariants of all of the Hamiltonian cycles in $K_7$. Because $f$ is a homeomorphism, for each Hamiltonian cycle $K \in C$ we have $a[f(K)] = a(K)$, independent of whether $f$ is orientation preserving or reversing (recall that a knot and its mirror image always have the same Arf invariant). For each $K \in C$, because $a(K) = a[f(K)]$ we have $a(K) + a[f(K)] = 0 \bmod 2$. There may be some Hamiltonian cycles $K$ such that $f(K) = K$, whereas for other cycles $K$, we have $f(K) \neq K$. Each Hamiltonian cycle $K$ such that $f(K) \neq K$ can be paired up with its image $f(K)$, and the sum of the Arf invariants of the pair will contribute nothing to $a$. By grouping together $a(K)$ and $a[f(K)]$ whenever $f(K) \neq K$, we see that $a = \sum_{K \in C} a(K) = \sum_{K = f(K)} a(K) \bmod 2$. That is, $a$ is equal to the mod 2 sum of the Arf invariants of just those Hamiltonian cycles such that $K = f(K)$. Because Theorem 5.2 shows that $a = 1$ for any embedding of $K_7$, there must be at least one Hamiltonian cycle $K$ such that $K = f(K)$. Now $f$ must either rotate or reflect the vertices around this $K$. Because $f$ induces an automorphism of order two on the vertices of $K_7$, each vertex is either interchanged with another vertex or is fixed by $f$. As there is an odd number of vertices, at least one vertex must be fixed by $f$. Also, not all of the vertices of $K_7$ can be fixed by $f$, because in this case the automorphism induced by $f$ would not be of order two; rather it would be of order one. However, this means that $f$ must reflect the vertices of $K$. Hence, at most, two vertices of $K$ are fixed by $f$. Because the total number of vertices on $K$ is odd, $f$ must fix precisely one vertex of $K$, and hence of $K_7$. □

Now we need some more definitions.

**Definition.** Let $G$ be a graph and let $\Phi$ be an automorphism of the vertices of $G$. Let $v$ be a vertex of $G$. Then the *orbit* of $v$ under $\Phi$ is the set of all vertices $w$ of $G$ such that for some $i \in \mathbb{N}$, we have $w = \Phi^i(v)$. The *order* of an orbit is the number of vertices in that orbit.

Observe that if a vertex $v$ is fixed by the automorphism $\Phi$, then the orbit of $v$ has order one. The reader should check that the orbits of vertices under an automorphism $\Phi$ partition the vertices into disjoint sets whose union is all of the vertices of $G$. We refer to these disjoint sets as the *orbits* of $\Phi$. The reader should also be able to prove that if the order of an orbit is $n$, then for every vertex $v$ in the orbit, $\Phi^n(v) = v$. If $h$ is a homeomorphism of $(\mathbb{R}^3, G)$ then $h$ induces an automorphism on the vertices of $G$, and we shall refer to the orbits of this induced automorphism as the *vertex orbits* of $h$. More generally, for any collection of vertices and edges $S$ of $G$ we define the *orbit* of $S$ under $h$ as the union of subsets of $G$ of the form $h^i(S)$ for some $i \in \mathbb{N}$. For example, if $S$ is a simple closed curve contained in $G$, then the orbit of $S$ is the collection of simple closed curves obtained by repeatedly applying $h$ to the curve $S$.

**Theorem 5.8.** *For every natural number* n, *the complete graph* $K_{4n+3}$ *is intrinsically chiral (Flapan & Weaver, 1992).*

*Proof.* Suppose that for some $n$, there is an embedding of $K_{4n+3}$ that is topologically achiral. Then there exists an orientation-reversing homeomorphism $h : (\mathbb{R}^3, K_{4n+3}) \to (\mathbb{R}^3, K_{4n+3})$. The homeomorphism $h$ induces some automorphism on the vertices of $K_{4n+3}$. This automorphism partitions the vertices into some number of orbits; let us say there are $p$ orbits. Let $n_1, \ldots, n_p$ denote the orders of all the orbits of the vertices of $K_{4n+3}$. Any $n_i$ that is not a power of two can be factored as $n_i = q_i 2^{m_i}$, where $q_i$ is an odd number. Now for each $i$, the homeomorphism $h^{q_i}$ induces an automorphism of order $2^{m_i}$ on the $n_i$ vertices in that orbit. Let $q$ denote the product of all the $q_i$. It follows that $q$ is odd, so $h^q$ is orientation reversing. Let $g = h^q$; then $g : (\mathbb{R}^3, K_{4n+3}) \to (\mathbb{R}^3, K_{4n+3})$ is an orientation-reversing homeomorphism, and the orders of all the vertex orbits of $g$ are powers of two (possibly including $2^0 = 1$).

Because the orders of the vertex orbits must add up to the total number of vertices $4n + 3$, and all the orbits have orders equal to a power of two, there must be at least one orbit of order one, and either at least one orbit of order two, or at least two additional orbits also of order one. In either case, it follows that at least three vertices are fixed by $g^2$. Now we consider four cases according to the orders of the vertex orbits of $g$.

*Case 1.* $g$ has at least one vertex orbit of order greater than two.

Hence there is an orbit of order $2^m$ for some $m \geqslant 2$. For this $m$, we see that $2^m$ is divisible by four. So there exists at least one vertex orbit $A$ of order four under the homeomorphism $g^{2^{m-2}} = g^{2^m/4}$. Now the homeomorphism $(g^{2^{m-2}})^2$ induces an automorphism of order two on the vertices in $A$. Consider the four vertices in $A$, together with the three vertices that we know are fixed by $g^2$. These seven vertices together with the edges between them form a subgraph that is an embedding of $K_7$. Now $(g^2)^{2^{m-2}} = (g^{2^{m-2}})^2$ is a homeomorphism of $(\mathbb{R}^3, K_7)$, which induces an automorphism of order two on this $K_7$. However, this contradicts Lemma 5.5 because $(g^2)^{2^{m-2}}$ fixes three of the vertices of $K_7$. Hence there is no vertex orbit of $g$ that has order greater than two.

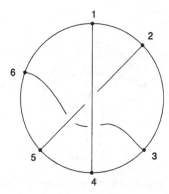

**Figure 5.12.** In Case 3, some embedding of this Möbius ladder would be setwise invariant under an orientation-reversing homeomorphism of $K_{4n+3}$

*Case 2.* $g$ has either one or two vertex orbits of order two.

Because we showed above that there are no orbits of order greater than two, those vertices of $K_{4n+3}$ that are not in orbits of order two, must be fixed by $g$. The complete graph $K_{4n+3}$ has at least seven vertices. Thus we can consider the subgraph whose vertices are contained in the one or two orbits of order two together with either five or three fixed vertices respectively, and together with all of the edges between pairs of these vertices. This subgraph is an embedding of $K_7$, and $g$ is a homeomorphism of $(\mathbb{R}^3, K_7)$, which induces an automorphism of order two on the vertices of this $K_7$. But as in Case 1, as more than one vertex is fixed by this automorphism, we contradict Lemma 5.5. Thus $g$ cannot have one or two vertex orbits of order two.

*Case 3.* $g$ has at least three vertex orbits of order two.

Label the vertices in three of the orbits of order two by the numbers $1, 2, \ldots, 6$, in such a way that the following three pairs of vertices each form an orbit of order two: $\langle 1, 4 \rangle$, $\langle 2, 5 \rangle$, and $\langle 3, 6 \rangle$. Observe that the edges $\overline{14}, \overline{25}$, and $\overline{36}$ are each setwise invariant under $g$. Also, $g$ interchanges the edge $\overline{12}$ with $\overline{45}$, the edge $\overline{23}$ with $\overline{56}$, and the edge $\overline{34}$ with $\overline{61}$. Let $M$ be the subgraph of $K_{4n+3}$ given by the vertices numbered one through six, together with the edges $\overline{14}, \overline{25}$, and $\overline{36}$, as well as the edges $\overline{12}, \overline{23}, \overline{34}, \overline{45}, \overline{56}, \overline{61}$. The edges $\overline{12}, \overline{23}, \overline{34}, \overline{45}, \overline{56}, \overline{61}$ form a simple closed curve, which we shall call $K$. Then $M$ is an embedding of a Möbius ladder with loop $K$ and rungs consisting of the edges $\overline{14}, \overline{25}$, and $\overline{36}$ (see Figure 5.12). In addition, $g$ is an orientation-reversing homeomorphism of $(\mathbb{R}^3, M)$ such that $g(K) = K$. We can easily extend $g$ to $S^3$ by defining $g(\infty) = \infty$. However, this contradicts Corollary 3.4, which showed that there

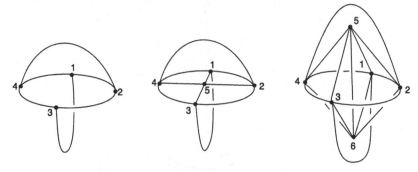

**Figure 5.13.** Rigidly achiral embeddings of $K_4$, $K_5$, and $K_6$

is no orientation-reversing homeomorphism of $S^3$ that takes a Möbius ladder with an odd number of rungs to itself and leaves the loop $K$ setwise invariant. Therefore, $g$ cannot have three or more orbits of order two.

*Case 4.* $g$ has no orbits of order two.

Because there were no vertex orbits of order a higher power of two, $g$ must fix every vertex in $K_{4n+3}$. But now we can choose any Möbius ladder with three rungs $M_3$ that is a subgraph of $K_{4n+3}$. Let $K$ be the loop of $M_3$. Because $g$ fixes every vertex of $M_3$, it is an orientation-reversing homeomorphism of $(\mathbb{R}^3, M_3)$ such that $g(K) = K$. As in Case 3, we obtain a contradiction to Corollary 3.4. Thus, $g$ cannot have zero orbits of order two.

Because none of the above cases can occur, we conclude that the orientation-reversing homeomorphism $h : (\mathbb{R}^3, K_{4n+3}) \to (\mathbb{R}^3, K_{4n+3})$ cannot exist. Hence $K_{4n+3}$ is intrinsically chiral. □

We shall now illustrate how the complete graphs that do not have $4n + 3$ vertices all have rigidly achiral embeddings. For $n < 5$, the complete graphs, $K_n$, are planar, and any planar embedding is rigidly achiral in $\mathbb{R}^3$. Figures 5.13 and 5.14 contain rigidly achiral embeddings of $K_4$, $K_5$, $K_6$, and $K_8$ in $\mathbb{R}^3$. We include $K_4$ in Figure 5.13 in order to illustrate the pattern, although $K_4$ also has a rigidly achiral embedding in the plane. Each graph in Figures 5.13 and 5.14 has an orientation-reversing homeomorphism obtained by rotating the graph by $90°$ about a central axis and then reflecting through a plane perpendicular to this axis. In the first three illustrations, this plane of reflection is perpendicular to the plane of the paper, whereas in the last illustration the plane of reflection is the plane of the paper.

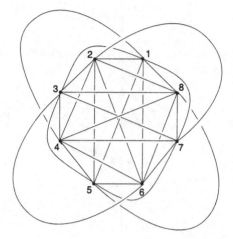

**Figure 5.14.** A rigidly achiral embedding of $K_8$

Although the construction is somewhat more complicated, we can create analogous embeddings for any $K_{4n}$, $K_{4n+1}$, and $K_{4n+2}$ as follows. We will first construct a rigidly achiral embedding of $K_{4n}$. Let $P$ be the $xy$ plane in $\mathbb{R}^3$ and let $C$ be the unit circle in $P$ centered at the origin. Also, let $h : \mathbb{R}^3 \to \mathbb{R}^3$ be the composition of a $90°$ rotation about the $z$ axis, and a reflection through the plane $P$. Choose the $4n$ vertices of $K_{4n}$ to be points on $C$ arranged in such a way that $f$ leaves these vertices setwise invariant. Label this set of vertices consecutively by the numbers $1, \ldots, 4n$. Now using straight line segments $a_1, \ldots, a_n$, connect the vertices $1, \ldots, n$ to a collection of $n$ distinct points on the positive $z$ axis. The homeomorphism $h^2$ fixes every point on the $z$ axis and takes each vertex $i$ on $C$ to its antipodal vertex $i + 2n$. For $i = 1, \ldots, n$, we define $e_i = a_i \cup h^2(a_i)$. Then each $e_i$ will be an arc from $i$ to $i + 2n$, and for each $i$ we have $h^2(e_i) = e_i$. Now we add to our construction of $K_{4n}$ the edges $e_1, \ldots, e_n$ as well as their images under $h$. Thus we have constructed disjoint edges between every pair of antipodal vertices on $C$. Figure 5.15 illustrates $C$ with twelve vertices and the edges $e_1$, $e_2$, and $e_3$.

Now choose nested ellipsoids $E_1, \ldots, E_n$, which all contain the circle $C$ but are otherwise disjoint, are each setwise invariant under $h$, and do not intersect the edges chosen above. Also for each $i = 1, \ldots, n$ and each $j$ such that $i < j < i + 2n$, let $F_{ij}$ denote a plane perpendicular to $P$ that contains vertices $i$ and $j$. For each such $i$ and $j$, we add an edge $f_{ij}$ (connecting vertex $i$ to vertex $j$), which is the arc in the intersection of $E_i$ and $F_{ij}$ that is on the side of $P$ where the $z$ axis is positive (see Figure 5.16). Note that for a fixed $i \leqslant n$ and any $j$ such that $i < j < i + 2n$, the edges $f_{ij}$ all lie in the ellipsoid $E_i$,

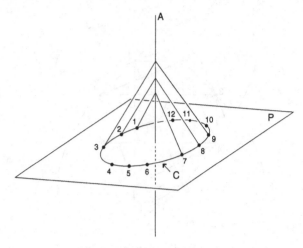

**Figure 5.15.** To create an achiral embedding of $K_{4n}$, we begin with this structure

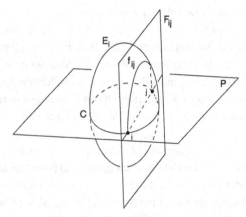

**Figure 5.16.** The edge connecting vertex $i$ to vertex $j$ is the arc in the intersection of $E_i$ and $F_{ij}$ that is on the side of the positive $z$ axis

but only intersect at the vertex $i$. Also for any $i \neq i'$ and any $j$ and $j'$, then $f_{ij}$ is contained in $E_i$, and $f_{i'j'}$ is contained in $E_{i'}$, so the edges $f_{ij}$ and $f_{i'j'}$ are disjoint except possibly at the vertex $j$ if $j = j'$. Because the edge $f_{ij}$ is in $E_i$ and all of the points in the interior of $f_{ij}$ have positive $z$ coordinates, the image of $f_{ij}$ under $h$ will be in $E_i$ and all of the points in its interior will have negative $z$ coordinates. Note that, because $i < j < i + 2n$ and $h^2$ rotates $E_i$ by $180°$, it follows that $h^2(f_{ij})$ is disjoint from $f_{ik}$ for all $k < i + 2n$. We now add to our graph the images of the edges $f_{ij}$ under $h$, $h^2$, and $h^3$. In this way we obtain edges between every pair of vertices in $K_{4n}$ (the reader should check this). By our construction, the set of edges will all be disjoint, and this embedding of $K_{4n}$ will be setwise invariant under the order-four orientation-reversing homeomorphism $h$ of $\mathbb{R}^3$.

To obtain a rigidly achiral embedding of $K_{4n+1}$, we start with the above rigidly achiral embedding of $K_{4n}$ and then we add a vertex at the center of the circle $C$ as well as straight line segments connecting this new vertex to the $4n$ vertices that are contained in the circle $C$. This embedding of $K_{4n+1}$ is also setwise invariant under the order-four orientation-reversing homeomorphism $h$ described above, and hence it is rigidly achiral.

To create a rigidly achiral embedding of $K_{4n+2}$, we start with the above embedding of $K_{4n}$, which is setwise invariant under the order-four orientation-reversing homeomorphism $h$. Then choose a point $x$ on the positive $z$ axis close enough to the origin so that the line segment on the $z$ axis, from $x$ to $h(x)$, does not intersect any of the edges of $K_{4n}$ or any of the ellipsoids $E_1, \ldots, E_n$. Let $x$ and $h(x)$ be the two new vertices, and include the line segment on the $z$ axis between them as an edge in our graph. Now add straight line segment edges from $x$ and $h(x)$ to each of the $4n$ vertices on the circle $C$. This embedding of $K_{4n+2}$ will be setwise invariant under the order-four orientation-reversing homeomorphism $h$ and hence will be rigidly achiral. The reader should check that the embeddings of Figures 5.12 and 5.13 correspond to the constructions that we have just given of rigidly achiral embeddings of $K_{4n}$, $K_{4n+1}$, and $K_{4n+2}$.

## Topological Stereoisomers

If we wish to understand the topological stereoisomers of a molecular graph, we have to pay attention to the labels of the vertices indicating which types of atoms these vertices represent. An automorphism of a molecular graph that permutes vertices with identical labels will have no apparent effect on the graph. For example, interchanging two oxygen atoms in a given molecular graph does not change the graph. However, if there is an automorphism of the labeled graph that permutes the positions of different types of atoms, performing this automorphism will create a new molecular graph. Recall from Chapter 1 that two molecular graphs, $G_1$ and $G_2$, are *rigid stereoisomers* if they are abstractly isomorphic as graphs with labeled vertices, but one is not rigidly superimposable on the other. In contrast, molecular graphs, $G_1$ and $G_2$, are said to be *topological stereoisomers* if they are abstractly isomorphic as graphs with labeled vertices, but one cannot be deformed to the other. This is equivalent to saying that $G_1$ and $G_2$ are abstractly isomorphic, but there is no orientation-preserving homeomorphism mapping $(\mathbb{R}^3, G_1)$ to $(\mathbb{R}^3, G_2)$.

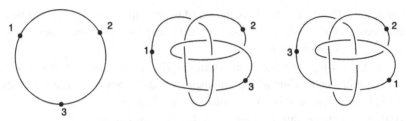

**Figure 5.17.** Different types of topological stereoisomers

In Chapter 1, we saw that two embeddings of a labeled graph may be rigid stereoisomers or topological stereoisomers because of their embeddings as unlabeled graphs, or because of the labeling of the vertices. Recall that the topological stereoisomers in Figure 5.17 illustrate this distinction. The first two graphs are topological stereoisomers because one is knotted and the other one is not. The second and third graphs are topological stereoisomers because of the labeling of the vertices. In particular, there is no orientation-preserving homeomorphism of $\mathbb{R}^3$, which sends the knot to itself and induces the automorphism (13) on the graph.

For a fixed embedding of a labeled graph, we would like to use the automorphisms of the vertices of the graph to create topological stereoisomers. Let $G$ be an embedded unlabeled graph in $\mathbb{R}^3$, and let $G_1$ and $G_2$ denote labelings of $G$ such that there is an automorphism $\Phi$ of the vertices of $G$ taking $G_1$ to $G_2$. Then $G_1$ and $G_2$ will be topological stereoisomers if there is no orientation-preserving homeomorphism $h : (\mathbb{R}^3, G) \to (\mathbb{R}^3, G)$ that induces $\Phi$ on the vertices of $G$.

We use the idea of automorphisms together with finite-order homeomorphisms to create a new type of stereoisomers, as follows.

**Definition.** Let $G$ be an embedded unlabeled graph in $\mathbb{R}^3$, and let $G_1$ and $G_2$ denote $G$ with two different labelings of the vertices. Then $G_1$ and $G_2$ are said to be *topologically rigid stereoisomers* if there is an automorphism $\Phi$ of the vertices of $G$ taking $G_1$ to $G_2$, but there is no finite-order orientation-preserving homeomorphism $h : (\mathbb{R}^3, G) \to (\mathbb{R}^3, G)$ that induces $\Phi$ on the vertices of $G$.

As always, there is an analogous definition for graphs embedded in $S^3$. Observe that if $G_1$ and $G_2$ are topologically rigid stereoisomers in $S^3$ then they are topologically rigid stereoisomers in $\mathbb{R}^3$, but not necessarily vice versa. To understand the relationship among our three types of stereoisomers, let us suppose that $G_1$ and $G_2$ are two labelings of an embedded graph $G$. If $G_1$ and

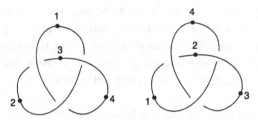

**Figure 5.18.** Topologically rigid stereoisomers that are not topological stereoisomers

$G_2$ are topological stereoisomers then they are necessarily topologically rigid stereoisomers, and if $G_1$ and $G_2$ are topologically rigid stereoisomers then they are necessarily rigid stereoisomers.

We illustrate the idea of topologically rigid stereoisomers with the two labeled trefoil knot graphs shown in Figure 5.18. The automorphism (1234) sends the first labeling to the second labeling. Suppose that $h : S^3 \rightarrow S^3$ has finite order $n$ and takes the trefoil knot $K$ to itself and induces the automorphism (1234) on the vertices of $K$. Then $h^{n/4}$ fixes the trefoil knot pointwise. A major result in topology is the proof of the Smith Conjecture, which states that if a nontrivial finite-order homeomorphism of $S^3$ fixes every point of a simple closed curve, then this simple closed curve must be unknotted (Morgan & Bass, 1984). Thus $h^{n/4}$ must be the identity map, so $h$ must have an order of four. However, there is no homeomorphism of $S^3$ that takes a trefoil knot to itself and has an order of four (see Hartley, 1981; Murasugi, 1971). It follows that there can be no finite-order homeomorphism of the trefoil knot in $S^3$ that induces the automorphism (1234) on $K$. Thus these two labeled trefoil knot graphs are topologically rigid stereoisomers in $S^3$ and hence in $\mathbb{R}^3$ as well. In contrast, they are not topological stereoisomers, because we can slither the first trefoil knot along itself to get an orientation-preserving homeomorphism of $S^3$ that sends the first labeling to the second labeling.

We should observe that the relationship between topologically rigid stereoisomers and topological stereoisomers is related to the relationship between embedded graphs that are rigidly achiral and those that are topologically achiral. Examples such as those illustrated in Figure 5.17 and Figure 5.18, of course, depend on the embedding of the graph in $\mathbb{R}^3$. If we embed a simple closed curve in the plane and put vertices on it, then any automorphism of the vertices can be realized by a finite-order homeomorphism of $\mathbb{R}^3$. In contrast, we will now give an example of an automorphism of the vertices of a complete graph that cannot be induced by a homeomorphism of $\mathbb{R}^3$, no matter how the graph is embedded. The fact that this automorphism cannot be induced by a homeomorphism of $\mathbb{R}^3$

is an intrinsic property of the graph, just as the properties of being intrinsically chiral, intrinsically knotted, or intrinsically linked are intrinsic properties of a graph. In order to prove that our automorphism cannot be induced by a homeomorphism, we will again make use of Theorem 5.1. Note that Theorem 5.9 (Flapan, 1989) could be just as well stated and proved in $S^3$.

**Theorem 5.9.** *Let $K_6$ be embedded in $\mathbb{R}^3$ with vertices labeled by the numerals* 1, 2, 3, 4, 5, 6. *Then, there is no homeomorphism of* h : $(\mathbb{R}^3, K_6) \rightarrow (\mathbb{R}^3, K_6)$ *that induces the automorphism* (1234).

*Proof.* Recall from the proof of Theorem 5.1 that any set of three vertices of $K_6$ represents a pair of disjoint loops in the graph. Because each pair of loops can be represented by two complementary sets of three numbers, there are a total of ten pairs of loops in $K_6$. We can write down the orbits of a pair of loops under the automorphism (1234) by specifying the images of one of the loops of the pair under the automorphism. These orbits are $\langle 123, 234, 341, 412 \rangle$, $\langle 125, 235, 345, 415 \rangle$, and $\langle 135, 245 \rangle$. Observe that each orbit contains an even number of pairs of loops.

Now suppose that there is an embedding of $K_6$ in $\mathbb{R}^3$ such that some homeomorphism $h : (\mathbb{R}^3, K_6) \rightarrow (\mathbb{R}^3, K_6)$ induces the automorphism (1234). Using the terminology of Theorem 5.1, for each pair of disjoint loops $A$ and $B$ we let $\omega(A, B)$ denote the mod 2 linking number of $A$ and $B$, and we let $\omega = \sum_{A,B} \omega(A, B)$ mod 2 where the sum is taken over all pairs of disjoint loops in $K_6$. Now because $h$ is a homeomorphism and the linking numbers are considered mod 2, $\omega(A, B) = \omega[h(A), h(B)]$ for every pair of loops $A$ and $B$ independent of whether $h$ is orientation preserving or reversing. Thus all of the pairs of loops in a given orbit will have the same mod 2 linking number. Because there is an even number of pairs of loops in each orbit, it follows that $\omega = \sum_{A,B} \omega(A, B)$ mod 2 = 0, contradicting the conclusion of Theorem 5.1 that $\omega = 1$ for every embedding of $K_6$ in $\mathbb{R}^3$. By this contradiction we conclude that there is no embedding of $K_6$ in $\mathbb{R}^3$ such that there is a homeomorphism $h : (\mathbb{R}^3, K_6) \rightarrow (\mathbb{R}^3, K_6)$ that induces the automorphism (1234).                                                                                        □

It follows from Theorem 5.9 that, given any embedding of $K_6$ in $\mathbb{R}^3$ with the vertices labeled 1, 2, 3, 4, 5, 6, if we perform the automorphism (1234) on the vertices we will obtain a new embedding of the labeled graph that is a topological stereoisomer of the original embedding. So, for example, Figure 5.19 illustrates two embeddings of the labeled graph $K_6$ that are topological stereoisomers.

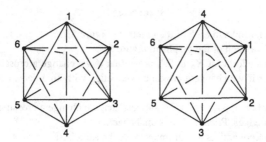

**Figure 5.19.** These labeled embeddings of $K_6$ are topological stereoisomers

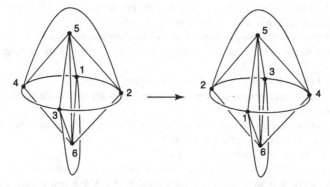

**Figure 5.20.** The automorphism (13)(24) is induced by a rotation

In general, we obtain topological stereoisomers if an automorphism cannot be induced by an orientation-preserving homeomorphism. Theorem 5.9 shows the stronger result that (1234) cannot be induced by any homeomorphism, whether orientation preserving or orientation reversing.

Note that if we apply the automorphism (1234) a second time, we will obtain an embedding of the labeled graph that is a topological stereoisomer of the embedding we got by applying the automorphism once. However, applying the automorphism (1234) twice will not necessarily give us a topological stereoisomer of our original embedding. In particular, applying the automorphism (1234) twice is the same as applying the automorphism (13)(24) once. Figure 5.20 illustrates an embedding of $K_6$ in which the automorphism (13)(24) is induced by a rotation of 180° about a vertical axis. Thus applying the automorphism (1234) once to the embedding in Figure 5.20 will give us a topological stereoisomer, but applying (1234) twice will give us an embedded graph that is rigidly superimposable on our original embedded graph.

## Exercises

1. Find nontrivial links in each of the graphs in Figure 5.7.
2. Compute $\omega$ for each of the graphs in Figure 5.7.
3. Given an embedding of $K_6$ in $\mathbb{R}^3$, prove that if we change a crossing of an edge with itself or if we change a crossing of an edge with an adjacent edge then $\omega$ is unchanged.
4. Given an embedding of $K_6$ in $\mathbb{R}^3$, suppose that we change a crossing between two nonadjacent edges. Prove that $\omega$ is unchanged.
5. Prove that a knot and its mirror image have the same Arf invariant.
6. Can we contract an edge of the embedding of $K_6$ shown in Figure 5.2 to obtain an embedded graph with no nontrivial link? Explain.
7. Does the embedding of $G = K_6 - \overline{14}$ in Figure 5.5 have the property that every simple closed curve in $G$ bounds a disk in the complement of $G$? Explain.
8. Prove that the graph $K_{3,3,1}$ is intrinsically linked.
9. Determine which of the graphs in Figure 5.7 can be obtained from $K_6$ by doing only one $\Delta Y$ move. What sequence of moves do we need to obtain the other graphs in Figure 5.7?
10. Prove Lemma 5.4 for intrinsically linked graphs.
11. Draw the thirteen graphs that can be obtained from $K_7$ by a finite number of $\Delta Y$ moves.
12. State and prove a lemma analogous to Lemma 5.5 for graphs that are minor minimal with respect to being intrinsically linked.
13. Find the two trefoil knots in Figure 5.11.
14. Find an embedding of $K_{3,3,1,1}$ that contains precisely one nontrivial knot.
15. Suppose that $\Phi$ is an automorphism of a graph $G$. Let $S$ denote the orbit of a vertex $v$ and let $T$ denote the orbit of a vertex $w$. Prove that if $S \neq T$ then $S$ and $T$ are disjoint.
16. Suppose that $\Phi$ is an automorphism of a graph $G$. Let $S$ be the orbit of a vertex $v$, and suppose that the order of $S$ is $n$. Prove that for every vertex $w \in S$, we have $\Phi^n(w) = w$.
17. Suppose that $h : (\mathbb{R}^3, G) \to (\mathbb{R}^3, G)$ is an orientation-reversing homeomorphism of finite order. Prove that there exists an orientation-reversing homeomorphism $f : (\mathbb{R}^3, G) \to (\mathbb{R}^3, G)$ whose order is a power of two.
18. Draw rigidly achiral embeddings of $K_9$ and $K_{10}$.
19. Two automorphisms $\Phi$ and $\lambda$ of a graph $G$ are said to be conjugate if there exists an automorphism $f$ of $G$ such that $\Phi = f \circ \lambda \circ f^{-1}$. How many nonconjugate automorphisms are there of $K_6$?
20. Consider all of the nonconjugate automorphisms of $K_6$ that are also not conjugate to (1234). For each such automorphism, find an embedding of $K_6$ in $S^3$ such that that automorphism is induced by a homeomorphism of $(S^3, K_6)$.
21. Consider the triangle $T$ with vertices 2, 6, and 5 in $K_7$. For each edge $e$ that is disjoint from $T$, find embeddings of $K_7 - e$ and $K_7/e$ that satisfy conditions 1 and 2, respectively, of Lemma 5.5.

# 6

# Symmetries of Embedded Graphs

In this chapter we will discuss different types of symmetries of embedded graphs. Results about symmetries will provide an easy method of proving that certain molecular graphs are intrinsically chiral. Specifically, by a *symmetry* of an embedded graph we shall mean a homeomorphism of $S^3$ or $\mathbb{R}^3$ that takes the graph to itself. In Chapter 4, an embedded graph was defined to be *rigidly achiral* in $S^3$ if there is an orientation-reversing homeomorphism $h : (S^3, G) \to (S^3, G)$ of finite order. By analogy with this definition, we shall define a *rigid symmetry* of an embedded graph $G$ as any finite-order homeomorphism $h : (S^3, G) \to (S^3, G)$. It is important to distinguish the concept of a rigid symmetry from that of a physically rigid motion of space. A physically rigid graph may have rotational symmetries, planar reflections, or combinations of these two types of symmetries, but no other types of symmetries. Rigid symmetries include these three types of rigid motions as well as other finite-order homeomorphisms that are not rigid motions. For example, Figure 6.1 illustrates an order-three homeomorphism that first deforms a trefoil to a symmetric position, then rotates it by 120°, and then deforms it back to its original shape.

   The concept of a symmetry of an embedded graph should also not be confused with that of an automorphism of an abstract graph, which is a map of the graph to itself, independent of any particular embedding of the graph in space. It is natural to wonder about the relationship between the automorphisms of a graph and the symmetries of embeddings of the graph. Every symmetry of an embedded graph induces an automorphism on the graph, but not every automorphism of a graph is induced by a symmetry of some embedding of the graph. For example, in Theorem 5.9, we showed that the automorphism (1234) cannot be induced by a symmetry of any embedding of $K_6$. Given an automorphism $\Phi$ of a graph $G$, we are interested in the following questions.

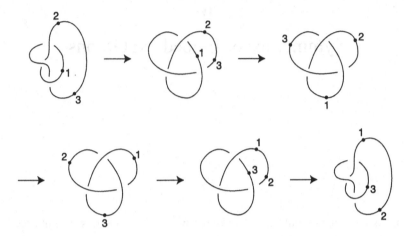

**Figure 6.1.** A rigid symmetry that is not a rigid motion

1. Is there an embedding of $G$ in $S^3$ and a homeomorphism $h:(S^3, G) \rightarrow$ $(S^3, G)$ that induces the automorphism $\Phi$?
2. Is there an embedding of $G$ in $S^3$ and a finite-order homeomorphism $h:(S^3, G) \rightarrow (S^3, G)$ that induces the automorphism $\Phi$?
3. Is there an embedding of $G$ in $S^3$ and an orientation-reversing homeomorphism $h:(S^3, G) \rightarrow (S^3, G)$ that induces the automorphism $\Phi$?

We can also ask analogous questions for graphs embedded in $\mathbb{R}^3$. The answers to questions 1 and 3 will be the same in $\mathbb{R}^3$ as they are in $S^3$, but for certain automorphisms the answer to question 2 may be different. Most of the results in this chapter are connected in one way or another with these three questions about the relationship between automorphisms and symmetries.

**Nonrigid Symmetries**

Just as there exist graphs that have topologically achiral embeddings but no rigidly achiral embeddings, there can exist graph automorphisms that can be induced by a homeomorphism for some embedding of the graph but cannot be induced by a finite-order homeomorphism of the graph, no matter how the graph is embedded in $\mathbb{R}^3$. Figure 6.2 illustrates a molecular graph $G$ with an automorphism $\varphi$ that cyclically permutes the vertices labeled M, I, and F, while fixing the rest of the vertices of the graph. For the given embedding in $\mathbb{R}^3$, this automorphism is induced by a homeomorphism that rotates the top of the graph by 120° while fixing the rest of the graph.

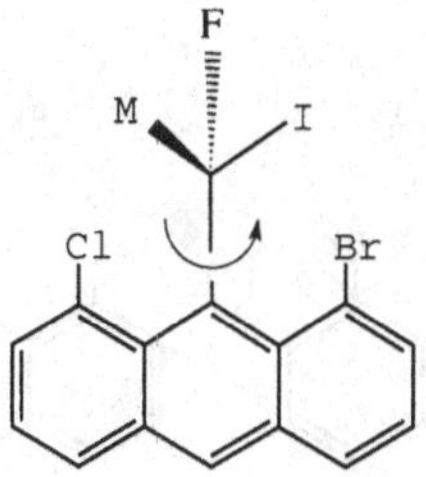

**Figure 6.2.** Automorphism $\varphi$ is induced by a homeomorphism but not by a finite-order homeomorphism

We prove as follows that there is no embedding of $G$ in $S^3$ such that the automorphism $\varphi$ is induced by a finite-order homeomorphism of $S^3$. Suppose there were such an embedding with a finite-order homeomorphism $h : (S^3, G) \to (S^3, G)$ that induces the automorphism $\varphi$. Then $h^2$ is not equal to the identity map because it moves the vertices M, I, and F. Also, independent of whether $h$ was orientation preserving or reversing, $h^2$ is necessarily an orientation-preserving finite-order homeomorphism. Because $h^2$ has finite order, its action on the graph $G$ is determined by its action on the vertices of $G$. It follows that $h^2$ pointwise fixes the three fused hexagons. This set of three hexagons is not homeomorphic to a subset of a circle. Thus the fixed-point set of $h^2$ is not the empty set or a set homeomorphic to a circle. This contradicts Smith's Theorem 4.1, because $h^2$ is orientation preserving. Hence no such finite-order homeomorphism can exist.

Because every finite-order homeomorphism of $\mathbb{R}^3$ can be extended to $S^3$, there is also no embedding of $G$ in $\mathbb{R}^3$ such that the automorphism $\varphi$ can be induced by a finite-order homeomorphism of $\mathbb{R}^3$. It follows that, for any embedding of the labeled graph $G$ in $\mathbb{R}^3$, performing the automorphism $\varphi$ gives another embedding that is a topologically rigid stereoisomer of the original one. In particular, the labeled graphs in Figure 6.3 are topologically rigid stereoisomers that are not topological stereoisomers. In contrast, the partial rotation illustrated in Figure 6.3 is chemically achievable, so these two graphs represent the same molecule, and hence are not chemical stereoisomers. In general, for molecular bond graphs that are topologically rigid stereoisomers but not topological stereoisomers, whether or not the molecules are chemical stereoisomers depends on to what extend the bonds can rotate around particular atoms.

## Rigid Symmetries of Three-Connected Graphs

We shall show that most complex graphs have the property that if a graph automorphism can be induced by a homeomorphism of $S^3$, for some embedding,

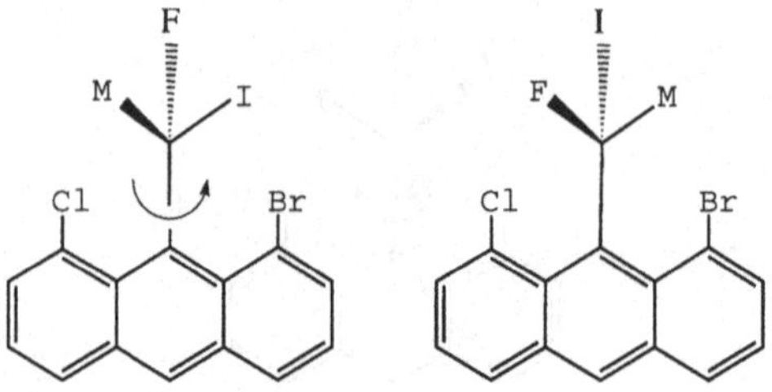

**Figure 6.3.** Topologically rigid stereoisomers that are not topological stereoisomers

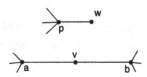

**Figure 6.4.** A three-connected graph cannot contain a vertex of valence one or two

then that same automorphism can be induced by a finite-order homeomorphism of $S^3$, for some (possibly different) embedding. We begin with the following definition.

**Definition.** A graph $G$ is said to be *n-connected* if at least $n$ vertices must be removed, together with the edges containing them, in order to disconnect $G$ or reduce $G$ to a single vertex.

Most of the graphs we have been working with are three connected. For example, any Möbius ladder is three connected and any complete graph on $n>3$ vertices is three connected (the reader should verify this). Also note that when we remove a vertex and the edges that contain it, we keep the vertices on the other end of any edges that are removed. We use this fact to prove, as follows, that any graph with a vertex of valence one or two cannot be three connected. Suppose that a graph $G$ contains a vertex $w$ of valence one. Then there is precisely one vertex $p$ adjacent to $w$. If we remove $p$ and the edges containing it, then either we are left with the single vertex $w$ or we have disconnected $G$ so that one component is $w$. If $G$ has a vertex $v$ that has valence two, then there are precisely two vertices $a$ and $b$ that are adjacent to $v$. If we remove both $a$ and $b$ and the edges containing them, then we either reduce the graph to the single vertex $v$ or we disconnect $G$ so that one component consists only of $v$. Thus in either case, $G$ is not three connected. Figure 6.4 illustrates these two possibilities.

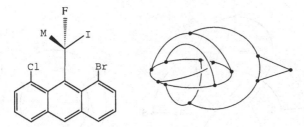

**Figure 6.5.** Graphs with nonrigid symmetries

**Figure 6.6.** A graph is not permitted to have these types of edges

Whether or not a graph is three connected will be especially important to us when considering symmetries in $S^3$. Intuitively, a graph is three connected if it does not have pieces that rotate independently. On the left side of Figure 6.5, we redraw the example of a graph with an automorphism that can be induced by a homeomorphism but, no matter how the graph is embedded in $S^3$, the automorphism cannot be induced by a finite-order homeomorphism. On the right side we redraw the example from Chapter 4 of a graph that is topologically achiral but no embedding of it is rigidly achiral in $S^3$. The reader can observe from these illustrations that neither of these graphs is three connected and both have pieces that rotate independently. We shall prove that graphs that are three connected do not have either of the properties of the graphs in Figure 6.5. Note that our results will be for graphs in $S^3$; the situation is more complicated for graphs in $\mathbb{R}^3$, as we shall see later in the chapter.

Before we state our theorem (Flapan, 1995) we recall that, by definition, graphs are not permitted to have a pair of edges with precisely the same two vertices and are not permitted to have any edges whose two endpoints share a single vertex. We illustrate the two forbidden types of edges in Figure 6.6. Neither of these types of edges occur in a molecular bond graph. Although there can be double bonds, these are considered as colored edges rather than as two independent edges that both have the same vertices.

**Theorem 6.1.** *Let* G *be a graph that is three connected, and let* $\Phi$ *be an automorphism of* G*. Suppose that* G *can be embedded in* $S^3$ *in such a way that* $\Phi$ *is induced by a homeomorphism* h : $(S^3, G) \to (S^3, G)$. *Then there exists an embedding of* G *in* $S^3$ *such that* $\Phi$ *is induced by a finite-order homeomorphism*

166                    6. Symmetries of Embedded Graphs

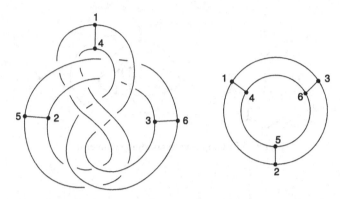

**Figure 6.7.** (123)(456) can be induced by a homeomorphism of the embedding on the left and by a finite-order homeomorphism of the embedding on the right

$f : (S^3, G) \to (S^3, G)$. *Furthermore, f is orientation reversing if and only if h is orientation reversing.*

To understand what Theorem 6.1 says, consider the embedded graph $G$ that is illustrated on the left of Figure 6.7. The reader should verify that this graph is three connected. Let $\Phi$ denote the automorphism (123)(456). This automorphism can be induced by the homeomorphism that slithers the figure eight knot along itself. Suppose that there is a finite-order homeomorphism $h$ of $S^3$ that induces the automorphism (123)(456) on $G$. Then $h^3$ pointwise fixes a figure eight knot. By the proof of the Smith Conjecture, there is no finite-order homeomorphism of $S^3$ that fixes a nontrivial knot pointwise (Morgan & Bass, 1984). It follows that $h$ has an order of three. However, it is known that there is no order-three homeomorphism of $S^3$ that takes a figure eight knot to itself (Hartley, 1981; Murasugi, 1971). Thus, for this particular embedding of the graph $G$, the automorphism $\Phi$ can be induced by a homeomorphism but not by a finite-order homeomorphism. Theorem 6.1 tells us that there is some other embedding of $G$ in $S^3$ such that $\Phi$ is induced by a finite-order homeomorphism of $S^3$. In fact, if we embed $G$ in a plane, as is illustrated on the right side of Figure 6.7, then $\Phi$ can be induced by a rotation of 120° about an axis perpendicular to the plane. In this particular example, there is a finite-order homeomorphism of $(\mathbb{R}^3, G)$ that induces $\Phi$. However, later in this chapter we shall see that this is not always the case.

We will not provide a detailed proof of Theorem 6.1 because the proof is quite technical and makes use of some sophisticated definitions and results in topology (for a detailed proof, see Flapan, 1995). However, even with the omission of some of the details, the proof is rather technical and it is not necessary to understand the proof of Theorem 6.1 in order to understand subsequent applications of it. So, on one hand, readers should feel free to skim the proof or to skip from here to the section on topological chirality of labeled graphs. On the other hand, for readers with some background in topology, the proof we present may provide a taste of the sort of argument one can construct by utilizing some of the recent deep results in topology, and hence further motivate these readers to continue their study of topology.

The basic idea of the proof is to start with an embedding of $G$ in $S^3$ that has a homeomorphism inducing $\Phi$. We consider a thickened version of $G$ in which the vertices and edges of the graph are replaced with balls and sticks. Then we cut open the complement in $S^3$ of this thickened version of $G$ along a special collection of tori and annuli. Using Thurston's Hyperbolization Theorem (Thurston, 1982), we shall define a finite-order homeomorphism $f$ on the component of the cut open three-manifold that is nearest to the boundary of the manifold. Finally, we re-embed $G$ in $S^3$ by sewing solid tori onto the torus boundaries of this component and extend $f$ within each of these solid tori. Readers who wish to read the proof given below should first become comfortable with the following definitions.

**Definition.** A surface $S$ with boundary is said to be *properly embedded* in a three-manifold $M$ if the boundary of $S$ is contained in the boundary of $M$ and the interior of $S$ is contained in the interior of $M$. When $S$ has no boundary then $S$ is *properly embedded* if $S$ is contained in the interior of $M$; and if $M$ has no boundary then any surface without boundary in $M$ is *properly embedded*.

For example, any torus embedded in $S^3$ is properly embedded because the torus has no boundary and $S^3$ has no boundary. A meridional disk is also properly embedded in a solid torus. A Möbius strip cannot be properly embedded in a three-dimensional ball since it is not possible for the boundary of the Möbius strip to lie in the boundary of the ball (readers with a background in algebraic topology should try to prove this).

**Definition.** A surface $S$ that is properly embedded in a three-manifold $M$ is said to be *incompressible* if every simple closed curve contained in $S$ that bounds a disk in $M$ also bounds a disk in $S$; otherwise the surface is said to be *compressible*.

For example, if $M$ is a solid torus, then its boundary is compressible because a meridian of the torus bounds a disk in the solid torus but not in the boundary of the solid torus. In contrast, if $M$ is the complement in $S^3$ of a knotted solid

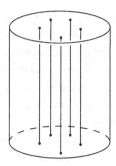

**Figure 6.8.** If we glue the top disk to the bottom disk with a $2\pi/5$ rotation, then these five line segments become a simple closed curve

torus together with its boundary, then it can be shown that the boundary of $M$ is incompressible. Because every simple closed curve in $S^3$ is either knotted or unknotted, any torus in $S^3$ is either the boundary of two solid tori or is the boundary of a solid torus and a knot complement.

We shall use the concept of a particular type of three-manifold called a *Seifert fibered space*, which we introduce below [see Jaco & Shalen (1979) for a more complete discussion]. We begin by writing the unit disk $D$ in polar coordinates as $D = \{(r, \theta)|r \in [0, 1]\}$. Then $D \times I = \{[(r, \theta), t] \,|r \in [0, 1], t \in [0, 1]\}$ is a solid cylinder. The circle $S^1$ can be obtained by gluing together the two ends of the line segment $I$. Thus we can obtain a solid torus $V = D \times S^1$ from $D \times I$ by gluing together the disks $D \times \{0\}$ and $D \times \{1\}$. However, there are many ways to glue these two disks together, which all give us the solid torus $V$. To understand the different ways, we start by letting $p$ and $q$ be integers with no common factors. Then we glue the two ends of the solid cylinder $D \times I$ together by gluing each point $[(r, \theta), 1]$ to the point $[r, \theta + (2\pi q/p), 0]$. Thus we are rotating the disk $D \times \{1\}$ by $2\pi q/p$ before we glue it to the disk $D \times \{0\}$. No matter what $p$ and $q$ are, for $x = (0, \theta)$, the two endpoints of the line segment $\{x\} \times I$ are glued together, creating a simple closed curve $f_x$ in the solid torus $V$. However, for $x = (r, \theta)$ where $r \neq 0$, the two endpoints of the line segment $\{x\} \times I$ will not be glued together. Rather, for each $i = 0, \ldots, p - 1$, and $x_i = [r, \theta + (2\pi i q/p)]$, the union of the line segments $f_x = \bigcup_{i=0}^{p-1} \{x_i\} \times I$ is a single simple closed curve in $V$. For example, in Figure 6.8, we will let $q = 1$, and $p = 5$, so that the top disk of the solid cylinder is glued to the bottom disk with a $2\pi/5$ rotation. In the solid torus $V$, the five line segments in the illustration become a single simple closed curve. For any gluing of the ends of the solid cylinder, we define a simple closed curve $f_x$ in $V$ for each $x \in D$. Furthermore $V = \bigcup_{x \in D} f_x$, and for each $x \in D$ and $y \in D$, either $f_x = f_y$ or $f_x$ and $f_y$ are disjoint.

Now $V$ is said to be a *fibered solid torus of type* $(p, q)$, and each $f_x$ is said to be a *fiber*.

Let $M$ be an orientable three-manifold. Suppose that $M$ can be written as a disjoint union of simple closed curves such that each curve is the core of a fibered solid torus whose fibers are the simple closed curves. Then we say that $M$ is a *Seifert fibered space*. It follows from classical results in topology that, for any Seifert fibered space with boundary, all of the boundary components must be tori.

The simplest example of a Seifert fibered space is just a fibered solid torus. We shall construct a slightly more complicated example as follows. Let $V$ be an unknotted solid torus in $S^3$. Then the complement of $V$ in $S^3$ is a solid torus $W$. Let $p$ and $q$ be nonzero integers other than 1 and $-1$, which share no common factors. Recall from Chapter 2 that a $(p,q)$-*torus knot* is a simple closed curve that goes $p$ times around a solid torus merdionally and $q$ times around the torus longitudinally. Let $K$ be a $(p, q)$-torus knot contained in the boundary of $V$. We can consider $V$ as the union of its core together with disjoint nested tori, which are parallel to the boundary of $V$. Each such torus (including the boundary of $V$) is the disjoint union of simple closed curves that go $p$ times around the torus merdionally and $q$ times around the torus longitudinally. Observe that a $(p, q)$ curve in the boundary of $V$ is also a $(q, p)$ curve in the boundary of the complementary solid torus $W$. Now $W$ is the union of its core together with disjoint nested tori, which are parallel to the boundary of $W$, and each of these tori is the disjoint union of simple closed curves that go $q$ times around the torus merdionally and $p$ times around the torus longitudinally. With some thought, the reader should see that $S^3$ is a Seifert fibered space where the fibers are the simple closed curves described above, together with the cores of the solid tori $V$ and $W$.

Now consider a solid tube $N$ whose core is the knot $K$, such that $N$ consists of the union of some of the simple closed curves described above, not including the cores of either of $V$ or $W$. We remove the interior of $N$ from $S^3$ to obtain the complement of a $(p, q)$-torus knot. This knot complement is also a Seifert fibered space, where the fibers are the simple closed curves described above together with the cores of $V$ and $W$. It can be shown that the only knots whose complements are Seifert fibered spaces are torus knots.

For the proof of Theorem 6.1 we need some more terminology, which we introduce below.

**Definition.** A *pared manifold* $(M, P)$ is an orientable three-manifold $M$ together with a specified family $P$ of disjoint incompressible annuli and tori in $\partial M$ (the symbol $\partial$ denotes the boundary of a set).

We use pared manifolds in order to keep track of which parts of the boundary of a thickened version of the graph correspond to edges of the graph. This, in turn, will allow us to keep track of the action induced by the automorphism $\Phi$ on the edges of the graph.

The following definition formalizes the idea of parallel surfaces from a topological point of view.

**Definition.** Let $T$ be a surface in a three-manifold $M$, and let $S$ be a component of the boundary of $M$. We say that $T$ is *parallel* to $S$ if the component of $M - T$ containing both $S$ and $T$ is homeomorphic to $T \times I$.

**Definition.** A pared manifold $(M, P)$ is said to be *simple* if it satisfies the following three conditions: (1) every two-dimensional sphere in $M$ bounds a ball and $\partial M - P$ is incompressible; (2) every incompressible torus in $M$ is parallel to a torus component of $P$; and (3) any annulus $A$ in $M$ with $\partial A$ contained in $\partial M - P$ is either compressible or is parallel to an annulus $A'$ in $\partial M$ with $\partial A' = \partial A$, and $A' \cap P$ consists of either zero or one annular component of $P$.

*Proof of Theorem 6.1 (With Some Details Omitted).* Suppose there exists an embedding of $G$ in $S^3$ such that $\Phi$ is induced by a homeomorphism $h : (S^3, G) \to (S^3, G)$. For each vertex $v_i$, let $N(v_i)$ denote a small ball with $v_i$ at the center, containing no other vertex of $G$, and for each edge $e_j$, let $N(e_j)$ denote a small solid tube around $e_j$ that contains no other points of the graph, and such that $N(v_i) \cap N(e_j)$ is either empty or a disk contained in $\partial N(v_i) \cap \partial N(e_j)$. For each $j$, let $b_j$ denote the cylindrical part of the boundary of $N(e_j)$. Thus each $b_j$ has two circles as its boundary components. Observe that $\Phi$ is uniquely determined by the automorphism that $h$ induces on the boundaries of $b_j$. Let $N(G)$ denote the union of all the balls $N(v_i)$ and the tubes $N(e_j)$ over all vertices and edges of the graph. Thus $N(G)$ is a thickened version of $G$. We can adjust $N(G)$ slightly, if necessary, so that $h[N(G)] = N(G)$. Let $M$ denote $S^3 - N(G)$ together with its boundary.

We apply Johannson's (1979) and Jaco–Shalen's (1979) Characteristic Torus Decomposition Theorem to $M$, in order to obtain a minimal family (i.e., a family containing as few tori as possible) $\tau$ of incompressible tori that separates $M$ into components $M_i$ such that for each $i$, either every incompressible torus in $M_i$ is parallel to a boundary component of $M_i$, or $M_i$ is a Seifert fibered space. Furthermore, by the Characteristic Torus Decomposition Theorem, any other minimal family that has these properties is ambient isotopic to $\tau$ by an isotopy that fixes $\partial M$. In particular, the family of tori $h(\tau)$ is ambient isotopic to $\tau$, by an isotopy $F$, which is fixed on $\partial M$. So, $F : M \times I \to M$ is a continuous function such that for each $t \in I$, $F(x, t)$ is a homeomorphism; $F(x, 0) = x$ for all $x \in M$; $F(h(\tau) \times \{1\}) = \tau$; and for all $x \in \partial M$, $F(x, t) = x$. Let $g : M \to M$ be defined by $g(x) = F(x, 1)$. Then $g$ is isotopic to the identity, and $g[h(\tau)] = \tau$. Now define $h' : (M, \tau) \to (M, \tau)$ by $h' = g \circ h$. Then $h'$ is isotopic to $h$ on $M$ by an isotopy that is fixed on $\partial M$.

Let $X$ be the component of $M - \tau$ containing $\partial M$. Then $h'(X) = X$. Furthermore, because $G$ is three connected, $G$ is connected and is not just a simple closed curve. Hence the boundary of $M$ is a connected surface that is not a torus. Each of the boundary components of a Seifert fibered space is a torus. Thus the component $X$ is not a Seifert fibered space, and so every incompressible torus in $X$ must be parallel to some boundary component of $X$.

Let $P_1$ denote the subset of $\partial M$ consisting of all of the $b_j$, and let $P_2$ denote those tori of $\tau$ that are contained in $\partial X$. Let $P = P_1 \cup P_2$. By using the hypothesis that $G$ is three connected, it can be shown that $\partial X - P$ is incompressible in $X$ (the reader should try to prove this). Hence, we can apply Johannson's (1979) and Jaco–Shalen's (1979) Characteristic Torus/Annulus Decomposition Theorem to the pared manifold $(X, P)$ in order to obtain a minimal family $F$ of incompressible tori and annuli in $X$ with boundaries in $\partial X - P$ that separates $(X, P)$ into components that either are simple or are Seifert fibered. Furthermore, the Characteristic Torus/Annulus Decomposition Theorem implies that any other minimal family of incompressible tori and annuli with these same properties is ambient isotopic to $F$ by an isotopy of $X$, which fixes each component of $P$. Because every incompressible torus in $X$ is already parallel to a component of $\partial X$, and the family $F$ is minimal, $F$ contains only incompressible annuli. Now using a similar argument to the one we used to obtain $h'$, we see that there is a homeomorphism $h'' : (X, F) \to (X, F)$ that is isotopic to $h'$ on $X$ and that has the property that $h''$ induces the same automorphism on the components of $P_1$ as $h$ did.

Let $N(F) = F \times [-1, 1]$, where our original $F$ is now represented by $F \times \{0\}$. Thus, $N(F)$ is a thickened version of $F$. We can choose $N(F)$ and $h''$ such that $h''[N(F)] = N(F)$. By again using the hypothesis that $G$ is three connected, together with some classical three-manifold topology, we can show that $X - N(F)$ has components $W, V_1, \ldots, V_n$, where each $V_i$ is homeomorphic to the product of a torus and an interval, $W$ is connected, and the pared manifold $\{W, W \cap [P \cup \partial N(F)]\}$ is simple. Furthermore, for each $i$, the component $V_i$ contains at most one component of $P_1$. Hence the automorphism that $h''$ induces on the components of $P_1$ is uniquely determined by the automorphism that $h''$ induces on the components of $W \cap [P \cup \partial N(F)]$.

Because $W$ is the only component of $X - N(F)$ which is not homeomorphic to the product of a torus with an interval, we can conclude that $h''(W) = W$.

Now by applying Thurston's Hyperbolization Theorem for Pared Manifolds (Thurston, 1982) and then Mostow's Rigidity Theorem (Mostow, 1973) and Waldhausen's Isotopy Theorem (Waldhausen, 1968), we obtain a finite-order homeomorphism $f$ of $\{W, W \cap [P \cup \partial N(F)]\}$ that is isotopic to $h''$ on $\{W, W \cap [P \cup \partial N(F)]\}$. In particular, $f$ induces the same automorphism as $h''$ on the components of $W \cap [P \cup \partial N(F)]$.

We shall explain as follows how to extend $f$ to a finite-order homeomorphism of $S^3$. Because $F \times \{1\}$ is contained in the boundary of $W$, $f$ is defined on $F \times \{1\}$. We now extend $f$ in a natural way to the product $N(F) = F \times [-1, 1]$ by defining $f(x, t) = f(x, 1)$ for every $x \in F$. Recall that each $V_i$ is a product of a torus with an interval. Thus one of the boundary components of $V_i$ is a torus boundary component of $X$. Also, the torus boundary components of $X$ are incompressible in $M$. It follows that each annulus in $\partial N(F) \cap V_i$ is attached to an annulus $A_i$ in $\partial M$ to form an incompressible torus in $M$. We have already defined $f$ on each annulus in $\partial N(F) \cap V_i$. Now we extend $f$ to each of the annuli $A_i$, in such a way that $f$ still has finite order. All of the tori in $\partial W$, as well as those tori made from the two annuli described above, are incompressible in $M$, and hence these tori bound knot complements in $M$. We now re-embed $W \cup \bigcup_{i=1}^{n} A_i$ in $S^3$ in such a way that each of these knot complements is replaced by a solid torus. Then we extend $f$ within each solid torus, and within each of the balls $N(e_j)$ and tubes $N(v_i)$ so that $f$ is defined on all of $S^3$ but still has finite order. Replacing the knot complements by solid tori also has the effect of re-embedding $G$ in $S^3$ in a simpler way. Finally, we conclude that, because $f$ induced the same automorphism as $h''$ on the components of $W \cap [P \cap \partial N(F)]$, it follows that $f$ induces the same automorphism as $h$ on the vertices of $G$. Furthermore, because $f$ is isotopic to $h''$ on $\{W, W \cap [P \cup \partial N(F)]\}$, $f$ is orientation reversing if and only if $h$ was orientation reversing.          $\Box$

## Topological Chirality of Labeled Graphs

Although the proof of Theorem 6.1 was quite difficult, it has two straightforward corollaries, which we will use in our analysis of topological stereoisomers. Note that both corollaries concern nonplanar graphs with no requirement that they be three connected. Both results are also about graphs in $\mathbb{R}^3$, rather than graphs that are embedded in $S^3$.

**Corollary 6.2.** *If* $G$ *is a nonplanar graph that is embedded in* $\mathbb{R}^3$, *then there is no orientation-reversing homeomorphism of* $(\mathbb{R}^3, G)$ *that induces the identity automorphism on the vertices of* $G$.

*Proof.* Suppose that there is an orientation-reversing homeomorphism $h:(\mathbb{R}^3, G) \to (\mathbb{R}^3, G)$ that induces the identity automorphism on the vertices of $G$. We can extend $h$ to an orientation-reversing homeomorphism of $S^3$ by defining $h(\infty) = \infty$. Because $G$ is nonplanar, Kuratowski's Theorem (Kuratowski, 1930) says that $G$ contains a subgraph that is homeomorphic to either a $K_{3,3}$ or a $K_5$ with the possible addition of some vertices of valence two. By removing these extra vertices, without removing any edges, we obtain a graph $H$ that is a subset of $G$ and that is $K_{3,3}$ or $K_5$. Because $K_{3,3}$ or $K_5$ are both three connected, so is the graph $H$. Also, $h$ is an orientation-reversing homeomorphism of $S^3$ that induces the identity automorphism on the vertices of $H$. Now we can apply Theorem 6.1 to get a (possibly different) embedding of $H$ in $S^3$ such that the identity automorphism on $H$ is induced by a finite-order orientation-reversing homeomorphism $f$ of $S^3$. But now the fixed-point set of $f$ contains the nonplanar graph $H$. Suppose that $H$ were embedded in a set homeomorphic to $S^2$. Because no embedding of a graph in $S^2 = \mathbb{R}^2 \cup \{\infty\}$ contains every point of $S^2$, we could choose some point in $S^2 - H$ to be the point $\infty$, giving us an embedding of $H$ in $\mathbb{R}^2$. As this is impossible, $H$ cannot be embedded in a set homeomorphic to $S^2$. By Smith's Theorem 4.1 the fixed-point set of any finite-order homeomorphism of $S^3$ must be contained in a set homeomorphic to $S^2$, unless $f$ is the identity map. However, because $f$ is orientation reversing, it cannot be the identity map. Thus no such orientation-reversing homeomorphism $h$ could exist. $\square$

We make a few simple observations about Corollary 6.2. First, observe that Corollary 6.2 is false for planar graphs because if we embed a graph in the plane then a reflection through that plane induces the identity automorphism on the graph. Furthermore, for any graph, the identity automorphism is induced by the identity map of $\mathbb{R}^3$, which is an orientation-preserving homeomorphism. So, by Corollary 6.2, the identity automorphism of a nonplanar graph is an example of an automorphism that cannot be induced by both an orientation-preserving homeomorphism and an orientation-reversing homeomorphism of $\mathbb{R}^3$. The following corollary shows that no automorphism of an embedded nonplanar graph can be induced by both an orientation-preserving and an orientation-reversing homeomorphism of $\mathbb{R}^3$.

**Corollary 6.3.** *Let G be a nonplanar graph that is embedded in $\mathbb{R}^3$, and let $\Phi$ be an automorphism of G. Then $\Phi$ cannot be induced by both an orientation-preserving homeomorphism of $(\mathbb{R}^3, G)$ and an orientation-reversing homeomorphism of $(\mathbb{R}^3, G)$.*

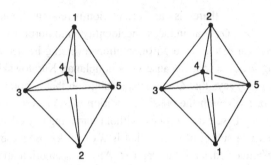

**Figure 6.9.** These labeled graphs are topological stereoisomers

*Proof.* Suppose that $\Phi$ could be induced by both an orientation-reversing home-omorphism $f$ of $(\mathbb{R}^3, G)$ and an orientation-preserving homeomorphism $g$ of $(\mathbb{R}^3, G)$. Let $h = f \circ g^{-1}$. Then $h$ is an orientation-reversing homeomorphism of $(\mathbb{R}^3, G)$ that induces the identity automorphism on the vertices of $G$. This contradicts Corollary 6.2.                                                                     $\square$

It is important in Corollary 6.3 that the embedding of $G$ in $\mathbb{R}^3$ is fixed. It is possible to have a nonplanar graph $G$ with an automorphism $\Phi$ such that, for some embedding of $G$ in $\mathbb{R}^3$, $\Phi$ is induced by an orientation-reversing homeomorphism of $\mathbb{R}^3$, and for a different embedding of $G$ in $\mathbb{R}^3$, $\Phi$ is induced by an orientation-preserving homeomorphism of $\mathbb{R}^3$ (see Exercise 15).

We can use Corollaries 6.2 and 6.3 to help us further analyze the topological stereoisomers of labeled graphs. Consider a nonplanar labeled graph $G$ embedded in $\mathbb{R}^3$. Let $\Phi$ be an automorphism that is nontrivial on the labeled graph. This means that $\Phi$ cannot simply interchange vertices that have identical labels. Rather, $\Phi$ must have a noticeable effect on at least some of the labeled vertices. Suppose that $\Phi$ is an automorphism of $G$ that can be induced by an orientation-reversing homeomorphism of $\mathbb{R}^3$. By Corollary 6.3, we know that, for this embedding of $G$ in $\mathbb{R}^3$, $\Phi$ cannot also be induced by an orientation-preserving homeomorphism. This means that if we perform $\Phi$ on the embedded graph $G$, then we will get a new embedding of the labeled graph that cannot be obtained by deforming the previous embedding. So, the new embedding will be a topological stereoisomer of the previous embedding. To illustrate this, consider the complete graph on five vertices, $K_5$, embedded as on the left of Figure 6.9. The automorphism (12) can be induced by reflecting the graph through the plane containing vertices 3, 4, and 5. Hence (12) cannot be induced by an orientation-preserving homeomorphism of $\mathbb{R}^3$. On the right in Figure 6.9, we have illustrated the embedding of $K_5$ after performing (12). Thus we cannot deform the embedding of the labeled $K_5$ that is on the left to get the labeled embedding of $K_5$ that is on the right of Figure 6.9. As these labeled graphs are mirror images of each other, we know our labeled graph is topologically chiral. Of course, as an unlabeled embedded $K_5$, this graph is topologically achiral.

Now suppose that $G$ is an embedded labeled graph in which two or more vertices have the same labels. We might have automorphisms $\varphi$ and $\Phi$, which are identical except at the vertices that have the same labels. There could be an orientation-reversing homeomorphism of $\mathbb{R}^3$ inducing $\Phi$ and an orientation-preserving homeomorphism of $\mathbb{R}^3$ inducing $\varphi$. Because $\Phi$ and $\varphi$ differed only on vertices with the same label, there would be no distinction between the labeled embedded graph we obtained by doing $\Phi$ and that we obtained by doing $\varphi$. Thus we could deform our original graph to get the one obtained by performing $\Phi$. For example, in Figure 6.9, if we imagine that vertices 3, 4, and 5 are all labeled with the same label, then we could induce the automorphism (12) by the deformation that turns the graph upside down. Also, if vertices 1 and 2 are both labeled with the same label and the rest of the vertices of $K_5$ are labeled as in Figure 6.9, then doing nothing to the graph would induce the same automorphism on the labels as performing a reflection through the horizontal plane containing vertices 3, 4, and 5. In both of these examples our embedded labeled $K_5$ would be topologically achiral.

Recall that a deformation of a molecular graph does not take a given type of atom (say an oxygen atom) into a different type of atom (say a carbon atom). This means that a labeled embedded graph is topologically achiral if and only if there is an orientation-reversing homeomorphism of $\mathbb{R}^3$ that takes the graph to itself and takes each labeled vertex to a vertex with the same label. For graphs in which every vertex is labeled distinctly, this means that the embedded graph is topologically achiral if and only if there is an orientation-reversing homeomorphism of $\mathbb{R}^3$ taking every vertex to itself. However, by Corollary 6.2 we know that if the graph is nonplanar then there can never exist such an orientation-reversing homeomorphism. Thus any nonplanar graph with all of its vertices labeled distinctly is topologically chiral, no matter how it is embedded in $\mathbb{R}^3$. So, by definition, every nonplanar graph with the vertices labeled distinctly is intrinsically chiral. A planar graph cannot be intrinsically chiral even if the vertices are labeled distinctly, because a planar embedding of it is necessarily topologically achiral. Hence, a graph with its vertices labeled distinctly is intrinsically chiral if and only if the graph is nonplanar.

From a topological point of view, it is simplest to consider graphs whose vertices either have no labels or are all labeled distinctly. However, molecular graphs have labeled vertices that are rarely all distinct. Thus, although we can still use Corollaries 6.2 and 6.3 to help us understand the topological stereoisomers of molecular graphs, we will have to look at the particular labeling of the atoms when we are evaluating the automorphisms of molecular graphs.

## Using Automorphisms to Show Intrinsic Chirality

We shall now present a corollary to Theorem 6.1 (Flapan, 1995) that enables us to prove that certain molecular and nonmolecular graphs are intrinsically chiral. Because a graph is intrinsically chiral in $S^3$ if and only if it is intrinsically chiral in $\mathbb{R}^3$, we shall prove the corollary in $S^3$ but apply it to graphs in $\mathbb{R}^3$.

**Corollary 6.4.** *Any nonplanar graph that has no order-two automorphism which is label-preserving is intrinsically chiral.*

*Proof.* Let $G$ be a nonplanar graph with no order-two automorphism which is label preserving. Suppose that there is an embedding of $G$ in $S^3$ such that there is an orientation-reversing homeomorphism $h : (S^3, G) \to (S^3, G)$ taking any labelled vertices to vertices with the same label. Then $h$ induces some automorphism $\Phi$ on the vertices of $G$. If $G$ had a label-preserving automorphism with an even order, then, by raising this automorphism to an appropriate power, we would obtain one of order two. So all of the label preserving automorphisms of $G$ must have an odd order. Hence for some odd number $p$, the automorphism $\Phi$ must have order $p$. Now $h^p$ is an orientation-reversing homeomorphism of $S^3$ which induces the identity automorphism on $G$. However, this contradicts Corollary 6.2. Hence, the graph $G$ must be intrinsically chiral. $\quad\square$

Note that every automorphism of an unlabeled graph is, by definition, label preserving. On the other hand, when we apply Corollary 6.4 to a molecular graph, we only have to consider automorphisms that send each atom to an atom of the same type.

To illustrate the application of Corollary 6.4 to molecular graphs we shall show that a ferrocenophan molecular graph and the Simmons–Paquette molecular graph are both intrinsically chiral. The Simmons–Paquette molecule is illustrated on the left of Figure 6.10, and the ferrocenophan molecule is illustrated on the right. Liang and Mislow (1994a) have also observed the intrinsic chirality of the ferrocenophan molecule and many other molecules. We saw in Chapter 1 that the molecular bond graph of the Simmons–Paquette molecule is a subdivision of a complete graph on five vertices and hence is nonplanar. Also, the ferrocenophan molecule contains a subgraph that is a subdivision of a complete graph on five vertices, although this is not as easy to see as in the Simmons–Paquette molecule. To make the $K_5$ easier to recognize in both graphs, the vertices of a $K_5$ subgraph have been numbered.

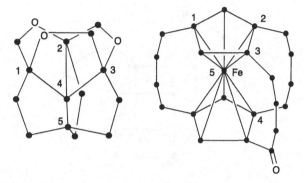

**Figure 6.10.** The graphs of the Simmons–Paquette molecule and ferrocenophan each contain a complete graph on five vertices

We prove, as follows, that the ferrocenophan graph is intrinsically chiral. First note that the central atom of the ferrocenophan molecule is iron and the rest of the atoms of the molecule are carbons, except for the oxygen, which is indicated in the figure. As usual, the hydrogen atoms have been omitted. In Exercise 6 the reader will prove that the ferrocenophan graph has no nontrivial automorphisms. Thus in particular, it has no automorphisms of order two. Because this graph is nonplanar, we immediately conclude by Corollary 6.4 that the graph is intrinsically chiral.

To prove that the Simmons–Paquette molecular graph has no order-two automorphisms requires slightly more work. All of the atoms of the Simmons–Paquette molecule are carbons, except for the oxygens indicated by Os in the diagram in Figure 6.10. Let $G$ denote the graph of the Simmons–Paquette molecule, and suppose that $G$ has an order-two automorphism $\Phi$ that takes oxygens to oxygens and carbons to carbons. Although the numbers that we have put on some of the vertices in Figure 6.10 have no chemical meaning, they are convenient for our proof. In particular, $\Phi$ must take each vertex that is numbered, to some numbered vertex, because the numbered vertices are the only vertices with valences of four. However, vertex 4 is the only valence four vertex whose shortest path to an oxygen contains exactly two edges, and vertex 5 is the only valence four vertex whose shortest path to an oxygen contains exactly three edges. Thus $\Phi$ must fix both vertex 4 and vertex 5. Now consider the action of $\Phi$ on vertices 1, 2, and 3. If $\Phi$ were to fix all three of these vertices then $\Phi$ would fix every vertex of $G$. This is not possible because $\Phi$ was assumed to have order two, whereas the identity automorphism has order one. Because of this assumption, $\Phi$ must fix one of the vertices 1, 2, and 3 and exchange the other two. Without loss of generality, suppose that $\Phi$ fixes vertex 1 and interchanges vertices 2 and 3. This implies that $\Phi$ interchanges a carbon and

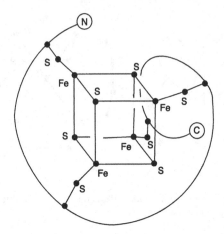

**Figure 6.11.** Chromatium high potential iron protein

an oxygen on either side of vertex 1. By this contradiction we conclude that no such order-two automorphism exists; hence the Simmons–Paquette molecule is intrinsically chiral.

Observe that although the Simmons–Paquette molecular graph has no order-two automorphisms, it does have an order-three automorphism that can be induced by a rotation of 120° about an axis containing vertices 4 and 5.

In addition to the Simmons–Paquette molecule and the ferrocenophan molecule, there are many other molecular graphs that contain $K_5$ or $K_{3,3}$ and hence are nonplanar. Some of these molecules can also be shown to be intrinsically chiral by using Corollary 6.4. Such nonplanar molecular graphs include some proteins as well as other synthetic structures (see Liang & Mislow, 1994a, 1994b). For example, the molecular graph of chromatium high potential iron protein (HiPIP) is illustrated in Figure 6.11. This protein is nonplanar because it contains a $K_{3,3}$, and it can also be shown to be intrinsically chiral by using Corollary 6.4 (see Exercise 9).

## Symmetries of Complete Graphs

In Theorem 5.9, we proved that the automorphism (1234) of the complete graph on six vertices cannot be induced by any homeomorphism of $\mathbb{R}^3$, no matter how $K_6$ is embedded in $\mathbb{R}^3$. It is natural to wonder whether there are other automorphisms of complete graphs, not conjugate to (1234), that cannot be induced by homeomorphisms of $\mathbb{R}^3$. It is easy to see that all of the automorphisms of $K_3$ can be induced by a rotation of $\mathbb{R}^3$ if we embed $K_3$ as an equilateral triangle. Similarly, every automorphism of $K_4$ can be induced by a rotation or a reflection

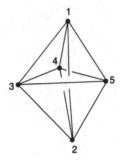

**Figure 6.12.** Every automorphism of $K_5$ is induced by a homeomorphism of this embedding

if we embed $K_4$ as a regular tetrahedron. The reader should check that every automorphism of $K_5$ can be induced by a homeomorphism of the embedding of $K_5$ illustrated in Figure 6.12, though not necessarily by a rotation, reflection, or rotation–reflection (see Exercise 4).

For $K_6$, it turns out that every automorphism, except those conjugate to (1234), can be induced by a homeomorphism of some embedding of $K_6$ (see Exercise 5). However, we can see as follows that not all of the automorphisms, which can be realized by some homeomorphism, can be induced by a homeomorphism of a single embedding of $K_6$ in $S^3$. Suppose that $K_6$ is embedded in $S^3$; we will show that there is some automorphism of $K_6$ that cannot be induced by a homeomorphism of this embedding but can be induced by a homeomorphism of a different embedding. Theorem 5.1 shows that for any embedding of $K_6$ in $\mathbb{R}^3$, the sum of the linking numbers of pairs of disjoint triangles $\omega = \sum_{A,B} \omega(A, B)$ mod $2 = 1$. It follows that for our particular embedding of $K_6$, there are an odd number of pairs of disjoint triangles with $\omega(A, B) = 1$, out of the total of ten pairs of disjoint triangles in $K_6$. This means that there must be at least one pair of triangles, say $\{A, B\}$, with $\omega(A, B) = 1$ and at least one pair of triangles, say $\{C, D\}$, with $\omega(C, D) = 0$. Thus, no homeomorphism of $S^3$ could take the pair of triangles $\{A, B\}$ to the pair of triangles $\{C, D\}$. Now without loss of generality we can label the vertices of $K_6$ so that $A = \triangle 123$, $B = \triangle 456$, $C = \triangle 423$, and $D = \triangle 156$. Then the automorphism (14) sends the pair $\{A, B\}$ to the pair $\{C, D\}$. Now since $\omega(A, B) = 1$ and $\omega(C, D) = 0$, there can be no homeomorphism of $S^3$ that induces the automorphism (14) for this embedding. However, the reader should check that there does exist an embedding such that (14) is induced by a homeomorphism of that embedding.

For example, in Figure 6.13, let $A = \triangle 123$, $B = \triangle 456$, $C = \triangle 423$, and $D = \triangle 156$. We can see that $\omega(A, B) = 1$ and $\omega(C, D) = 0$, and the

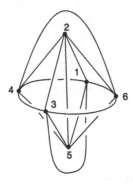

**Figure 6.13.** Automorphism (14) cannot be induced by a homeomorphism of this embedding

automorphism (14) sends the pair $\{A, B\}$ to the pair $\{C, D\}$. So, by the above argument, there is no homeomorphism of $S^3$ that induces the automorphism (14) for this embedding. Of course, (13) can be induced by a homeomorphism of this embedding and (14) can be induced by a homeomorphism of a different embedding of $K_6$ in $S^3$.

For any $n > 6$, $K_n$ will contain $K_6$, and hence the automorphism (1234) also cannot be induced by a homeomorphism of $(S^3, K_n)$, no matter how $K_n$ is embedded in $S^3$. Evaluating which other automorphisms of $K_n$ can be induced by a homeomorphism seems to get increasingly complicated as the number of vertices increases. However, by using Theorem 6.1, we shall be able to characterize which automorphisms of any complete graph with at least seven vertices can be realized by a homeomorphism of some embedding of the graph in $\mathbb{R}^3$. In order to prove that our list of automorphisms is exhaustive, we shall need seven lemmas, which we state and prove below. Each lemma assumes that $\Phi$ is an automorphism of $K_n$ that is induced by a finite-order homeomorphism $f$ for some embedding of $K_n$ in $S^3$, and then uses Smith's Theorem 4.1 to draw conclusions about either $\Phi$ or $f$. All of the lemmas are taken from Flapan (1995).

**Lemma 6.5.** *Suppose that* $n > 6$ *and* $\Phi$ *is an automorphism of* $K_n$ *that is induced by a homeomorphism* $f : (S^3, K_n) \to (S^3, K_n)$ *with an order of* $m$, *for some* $m > 1$. *Then the following statements are true :* (1) *the fixed-point set of* $f$ *is not homeomorphic to a two-dimensional sphere;* (2) $\Phi$ *fixes at most three vertices.*

*Proof.* Suppose that the fixed point set $Z$ of $f$ is homeomorphic to a two-dimensional sphere. This implies that $f^2$ is orientation preserving and fixes a set containing at least $Z$. By Smith's Theorem 4.1 this is impossible unless $f$ had order two. If $\Phi$ fixed at least five vertices, then $f$ would pointwise fix a

$K_5$ that is a subgraph of $K_n$. Because $K_5$ is nonplanar, it cannot be entirely contained in the fixed point set $Z$. Thus $\Phi$ fixes at most four vertices of $K_n$. Because $n > 6$ and $f$ has order two, this means that there must be at least two vertex orbits that have order two. Let these two orbits of order two be $\langle v_1, w_1 \rangle$ and $\langle v_2, w_2 \rangle$, and let the components of $S^3 - Z$ be denoted by $A$ and $B$. Because $f$ interchanges $A$ and $B$, we can assume, without loss of generality, that $v_1$ and $v_2$ are in $A$, whereas $w_1$ and $w_2$ are in $B$. Consider the edge $e = \overline{v_1 w_2}$. Because $e$ has one endpoint in $A$ and the other endpoint in $B$, there must be some point where $e$ intersects $Z$. However, this means that $f$ fixes this one point on the edge $e$. Hence $f(e) = e$, because $f$ takes edges to edges. Thus $f(v_1) = w_2$, which is contrary to our assumption that $f(v_1) = w_1$. So $f$ cannot have a fixed point set that is homeomorphic to a two-dimensional sphere. This proves part 1.

If $\Phi$ fixes at least four vertices, then $f$ will pointwise fix a $K_4$ that is a subgraph of $K_n$. However, by Smith's Theorem 4.1, because the fixed-point set of $f$ is not homeomorphic to a two-dimensional sphere, it must either be the empty set, a set of two points, or a set that is homeomorphic to a circle. Because $K_4$ cannot be contained in any of these sets, $\Phi$ cannot fix more than three vertices. This proves part 2. □

**Lemma 6.6.** *Suppose that* $n > 6$ *and* $\Phi$ *is an automorphism of* $K_n$ *that is induced by a homeomorphism* $f : (S^3, K_n) \to (S^3, K_n)$ *of order two. Then* $\Phi$ *has at most two fixed vertices and* $f$ *is orientation preserving.*

*Proof.* Suppose that $\Phi$ fixes three vertices. Then $f$ must pointwise fix the edges between these vertices. Because this set of three vertices and three edges is homeomorphic to a circle, by Smith's Theorem 4.1 and Lemma 6.5, $f$ can fix no other points of $S^3$. Because $n > 6$ there must be at least four vertices that are not fixed by $\Phi$. Hence there are at least two orbits of order two, because the order of $f$ is two. Let $\langle v, w \rangle$ denote an orbit of order two. Let $e = \overline{vw}$; then $f(e) = e$. Thus $f$ fixes a point of $e$, but this contradicts our above statement about the fixed point set of $f$. Therefore, there can be at most two vertices fixed by $\Phi$.

Because $n > 6$, and $\Phi$ fixes no more than two vertices, there must exist at least three vertex orbits of order two, say $\langle v_1, w_1 \rangle$, $\langle v_2, w_2 \rangle$, and $\langle v_3, w_3 \rangle$. Let $e_1 = \overline{v_1 w_1}$, $e_2 = \overline{v_2 w_2}$, and $e_3 = \overline{v_3 w_3}$. Then $f(e_i) = e_i$ for each $i$. Therefore, $f$ must fix a point on each of the edges $e_i$. We saw in Lemma 6.5 that the fixed point set of $f$ cannot be homeomorphic to a two-dimensional sphere. So by Smith's Theorem 4.1, if $f$ is orientation reversing then the fixed point set of $f$ is just two points. As $f$ fixes one point on each of at least three edges, $f$ cannot be orientation reversing. Thus $f$ is orientation preserving. □

**Lemma 6.7.** *Suppose that* n > 6 *and* $\Phi$ *is an automorphism of* $K_n$ *that is induced by a homeomorphism* $f:(S^3, K_n) \to (S^3, K_n)$ *of order* m, *for some* m $\in$ $\mathbb{N}$. *Then the order of* $\Phi$ *is also* m.

*Proof.* Suppose that the order of $\Phi$ is some $p \neq m$. Because $f^m$ is the identity, $p$ must divide $m$. Now $\Phi^p$ is induced by the finite-order homeomorphism $f^p$, which is not the identity. But $\Phi^p$ fixes every vertex, and there are at least seven vertices, so this contradicts Lemma 6.5. Therefore, the order of $\Phi$ must also be $m$. $\square$

Note that we say that an orbit is *trivial* if the orbit has order one, and otherwise we say that the orbit is *nontrivial*.

**Lemma 6.8.** *Suppose that* n > 6 *and* $\Phi$ *is an automorphism of* $K_n$ *that is induced by a homeomorphism* $f:(S^3, K_n) \to (S^3, K_n)$ *of order* m, *for some* m > 3. *Then the following statements are true :* (1) *all of the nontrivial orbits of* $\Phi$ *are of order* 2, 3, *or* m; (2) $\Phi$ *has at most one orbit of order* 3, *and if* $\Phi$ *has an orbit of order* 3 *then* $\Phi$ *has no fixed vertices;* (3) $\Phi$ *has at most one orbit of order* 2; *and* (4) $\Phi$ *has at least one orbit of order* m.

*Proof.* Any vertex that is fixed by $\Phi$ is also fixed by $\Phi$ raised to any power. Thus for any $p > 1$, the automorphism $\Phi^p$ fixes all the vertices that were fixed by $\Phi$ as well as all the vertices in any orbits of order $p$. Also $\Phi^p$ is induced by the finite-order homeomorphism $f^p$. If $p < m$, then the automorphism $\Phi^p$ is not the identity by Lemma 6.7. Thus, by Lemma 6.5, the automorphism $\Phi^p$ cannot fix more than three vertices. If $3 < p < m$ and there were an orbit of order $p$, then $\Phi^p$ would fix more than three vertices. If $\Phi$ had more than one orbit of order 3 or an orbit of order 3 and some fixed vertex, then $\Phi^3$ would also fix more than three vertices. By the same argument, $\Phi$ cannot have more than one orbit of order 2, or an orbit of order two and more than one fixed vertex.

Now suppose that $\Phi$ has no orbit of order $m$. It follows from what we have just proved that every nontrivial orbit must be of order 2 or 3. Also, as we saw above, there can be no more than one orbit of order 2 together with no more than one orbit of order 3, and no more than one fixed vertex. However, this means that $K_n$ contains no more than six vertices, which is contrary to hypothesis. Hence $\Phi$ must have at least one orbit of order $m$. $\square$

**Lemma 6.9.** *Suppose that* n > 6 *and* $\Phi$ *is an automorphism of* $K_n$ *that is induced by an orientation-preserving homeomorphism* $f:(S^3, K_n) \to (S^3, K_n)$

*of order* m, *for some* m *that is even and* m $> 2$. *Then* f *has no fixed points and* $\Phi$ *has no orbits of order* 2.

*Proof.* Suppose that the fixed point set of $f$ is not empty. Then by Smith's Theorem 4.1, because $f$ is orientation preserving, $f$ has fixed-point set $J$ that is homeomorphic to a circle. Now for any $r < m$, the homeomorphism $f^r$ is orientation preserving, of finite order, and fixes every point of $J$. Thus by Smith's Theorem 4.1, for every $r < m$, the fixed-point set of $f^r$ is also precisely the set $J$.

Because $m > 2$ is even, we know that $m > 3$, so it follows from Lemma 6.8 that $\Phi$ has at least one orbit of order $m$. Let this orbit be given by $\langle v_1, v_2, \ldots, v_m \rangle$, where $\Phi(v_i) = v_{i+1}$ for each $i = 1, \ldots, m$ and the subscripts are considered mod $m$. Because $m$ is assumed to be even, $r = m/2$ is a natural number. Now $\Phi^r(v_1) = v_{r+1}$ and $\Phi^r(v_{r+1}) = v_1$. Let $e = \overline{v_1 v_{r+1}}$; then $f^r(e) = e$. Thus some point of $e$ must be fixed by $f^r$. Because $f$ has the same fixed-point set as $f^r$, this implies that a point of $e$ must also be fixed by $f$. Because $f$ takes edges to edges, we now must have $f(e) = e$. However, this means that $\Phi(v_1) = v_{r+1}$ and $m > 2$ means that $r > 1$, which contradicts our assumption that $\Phi(v_i) = v_{i+1}$ for each $i$. This contradiction implies that $f$ has no fixed points.

Suppose now that $\Phi$ has an orbit of order two which we shall denote by $\langle v, w \rangle$. Let $a = \overline{vw}$; then $f(a) = a$. Thus $f$ would fix a point of the edge $a$. Because $f$ fixes no points, $\Phi$ must have no orbits of order two. $\square$

**Lemma 6.10.** *Suppose that* n $> 6$ *and* $\Phi$ *is an automorphism of* $K_n$ *that is induced by an orientation-reversing homeomorphism* f$: (S^3, K_n) \to (S^3, K_n)$ *of order* m, *for some* m $\in \mathbb{N}$. *Then* m $= 4$ *and* $\Phi$ *fixes at most one vertex. Furthermore, if* $\Phi$ *has an orbit of order* 2, *then* $\Phi$ *fixes no vertices.*

*Proof.* Performing an orientation-reversing homeomorphism an odd number of times will yield an orientation-reversing homeomorphism, so the order of any finite-order orientation-reversing homeomorphism must be an even number. Thus $r = m/2$ is a natural number. Now, $f^r$ has order two, so by Lemma 6.6, $f^r$ must be orientation preserving. Hence $r$ must also be even. Because $f$ is orientation reversing, it has a nonempty fixed-point set by Smith's Theorem 4.1. The fixed-point set of $f^2$ includes that of $f$. Thus $f^2$ has a nonempty fixed-point set, is orientation preserving, and has an even order $r$. Unless $r = 2$ we will get a contradiction to Lemma 6.9. Therefore $r = 2$ and hence $m = 4$. By Lemma 6.5, the fixed point set of $f$ is not homeomorphic to a two-dimensional sphere. Because $f$ is orientation reversing, it follows from Smith's Theorem 4.1 that the fixed-point set of $f$ is precisely two points. If $\Phi$ fixed two vertices of $K_n$ then $f$ would fix every point of the edge joining these two vertices. Thus, $\Phi$ fixes at most one vertex of $K_n$.

Now suppose that $\Phi$ has an orbit of order 2 and $\Phi$ fixes some vertex. Because $m = 4$, the homeomorphism $f^2$ has order 2. But $f^2$ induces the automorphism $\Phi^2$, which fixes at least three vertices. This contradicts Lemma 6.6. Therefore, if $\Phi$ has an orbit of order 2, then $\Phi$ fixes no vertices.  □

**Lemma 6.11.** *Suppose that* n > 6 *and* $\Phi$ *is an automorphism of* $K_n$ *that is induced by a homeomorphism* $f:(S^3, K_n) \to (S^3, K_n)$ *of order* m, *for some even* m. *Then* $\Phi$ *has no orbits of order three.*

*Proof.* Because $m$ is even, the homeomorphism $f^3$ that induces $\Phi^3$ also has an even order. If the order of $f^3$ is two, then by Lemma 6.6, $\Phi^3$ fixes at most two vertices. If the order of $f^3$ is more than two and $f^3$ is orientation preserving, then by Lemma 6.9, $\Phi^3$ fixes no vertices. Finally, if $f^3$ is orientation reversing then $\Phi^3$ fixes at most one vertex. Now suppose that $\Phi$ has an orbit of order three. Then $\Phi^3$ fixes at least the three vertices that were in this orbit, contradicting each of the above possibilities. Thus $\Phi$ cannot have any orbits of order three.  □

Now we are ready to state and prove our theorem characterizing which automorphisms of complete graphs with at least seven vertices can be induced by a homeomorphism of $\mathbb{R}^3$ for some embedding of the graph in $\mathbb{R}^3$. Note that Theorem 6.12 (Flapan, 1995) does not assume that the homeomorphism $h$ has finite order. Instead, as complete graphs with at least seven vertices are three connected, we shall use Theorem 6.1 in order to prove Theorem 6.12.

**Theorem 6.12.** *Let* $K_n$ *be a complete graph on* n > 6 *vertices, and let* $\Phi$ *be an automorphism of* $K_n$ *of order* m, *for some* m $\in \mathbb{N}$. *Then there is an embedding of* $K_n$ *in* $\mathbb{R}^3$ *such that* $\Phi$ *is induced by an orientation-preserving homeomorphism* $h:(\mathbb{R}^3, K_n) \to (\mathbb{R}^3, K_n)$ *if and only if* $\Phi$ *satisfies one of the following conditions:* (1) m *is even,* m > 2, *and all the orbits of* $\Phi$ *have an order of* m; (2) m = 2, *all the nontrivial orbits of* $\Phi$ *have an order of* m, *and there are at most two fixed vertices;* (3) m *is odd, all the nontrivial orbits of* $\Phi$ *have an order of* m, *and there are at most three fixed vertices; and* (4) m *is an odd multiple of three,* m > 3, *and all the orbits of* $\Phi$ *have an order of* m *except one, which has an order of three. Furthermore, there is an embedding of* $K_n$ *in* $\mathbb{R}^3$ *such that* $\Phi$ *is induced by an orientation-reversing homeomorphism* $h:(\mathbb{R}^3, K_n) \to (\mathbb{R}^3, K_n)$ *if and only if* $\Phi$ *satisfies the following condition:* (5) m = 4, *either all the nontrivial orbits of* $\Phi$ *have an order of four and there is at most one fixed vertex, or all the orbits of* $\Phi$ *have an order of four except one orbit, which has an order of two.*

In order to prove that all of the automorphisms listed in Theorem 6.12 can be induced by a homeomorphism for some embedding of $K_n$ in $\mathbb{R}^3$, we must construct an appropriate embedding for each type of automorphism. This is done in Flapan (1995). We will not include the constructions here because they are quite tedious. Below we prove that any automorphism induced by a homeomorphism satisfies one of the five conditions.

*Proof.* Suppose that $\Phi$ is an automorphism of $K_n$ that is induced by a homeomorphism $h$ of $\mathbb{R}^3$ for some embedding of $K_n$ in $\mathbb{R}^3$. We shall prove that $\Phi$ satisfies one of the conditions of our theorem. We begin by extending $h$ to $S^3$ by defining $h(\infty) = \infty$. Because $K_n$ is three connected, we can now apply Theorem 6.1 to re-embed $K_n$ in $S^3$ in such a way that $\Phi$ is induced by a finite-order homeomorphism $f : (S^3, K_n) \to (S^3, K_n)$, and $f$ is orientation reversing if and only if $h$ was.

Now we will apply our various lemmas to show that $\Phi$ satisfies one of our conditions. By Lemma 6.7, since $\Phi$ has an order of $m$, the homeomorphism $f$ must also have an order of $m$. Now by Lemma 6.8, all of the nontrivial orbits of $\Phi$ must have order two, three, or $m$, and if $m > 3$ there is at most one orbit of order 2 and at most one of order 3. Furthermore, according to Lemma 6.5, $\Phi$ has at most three fixed vertices.

We begin by considering the case in which $h$, and hence $f$, is orientation preserving. Suppose first that $m$ is odd. In this case, $\Phi$ cannot have an orbit of order two because the order of each orbit must divide $m$. If $\Phi$ does not have an orbit of order 3 or if $m = 3$, then $\Phi$ satisfies condition 3. Suppose that $m > 3$ and $\Phi$ has an orbit of order 3. Then $m$ is an odd multiple of three, and by Lemma 6.8, $\Phi$ has no fixed vertices. Hence $\Phi$ satisfies condition 4.

Now suppose that $m$ is even. By Lemma 6.11, $\Phi$ has no orbits of order 3. If $m > 2$, because we are assuming that $f$ is orientation preserving, we can apply Lemma 6.9 to conclude that $\Phi$ has no orbits of order 2 and $\Phi$ has no fixed vertices. In this case, $\Phi$ satisfies condition 1. If $m = 2$, then all orbits must have orders of two, because the order of each orbit must divide $m$. Also, according to Lemma 6.6, $\Phi$ has at most two fixed vertices. Hence, now $\Phi$ satisfies condition 2. In summary, if $h$ is orientation preserving then $\Phi$ must satisfy one of conditions 1, 2, 3, or 4.

Next we consider the case in which $h$, and hence $f$, is orientation reversing. By Lemma 6.10, we must have that $m = 4$ and $\Phi$ fixes at most one vertex. On one hand, if $\Phi$ has no orbit of order 2 then $\Phi$ satisfies condition 5. On the other hand, if $\Phi$ has an orbit of order 2, then by Lemma 6.8, $\Phi$ has at most one orbit of order 2, and by Lemma 6.10, $\Phi$ has no fixed vertices. So, again $\Phi$ satisfies condition 5. Therefore, if $h$ is orientation reversing then $\Phi$ must satisfy condition 5. $\quad\square$

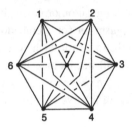

**Figure 6.14.** Automorphism (135)(246) of $K_7$ can be induced by a rotation

In Chapter 5, we showed, for every $n \in \mathbb{N}$, how to embed $K_{4n}$, $K_{4n+1}$, and $K_{4n+2}$ in $\mathbb{R}^3$ in such a way that they are topologically achiral, and we proved that for every $n \in \mathbb{N}$, $K_{4n+3}$ is intrinsically chiral. No complete graph with an automorphism satisfying condition 5 of Theorem 6.12 could have $4n + 3$ vertices. So Theorem 6.12 provides another proof of Theorem 5.6. The homeomorphisms for the embedded graphs $K_{4n}$, $K_{4n+1}$, and $K_{4n+2}$ that were given in Chapter 5 will induce the automorphisms satisfying condition 5 of Theorem 6.12. For each of the other automorphisms described by Theorem 6.12, a similar type of embedding can be constructed that has a homeomorphism inducing that automorphism. To give the flavor of these embeddings, we present a specific example.

The automorphism (135)(246) of $K_7$ satisfies condition 3 of Theorem 6.12. This automorphism can be induced by a homeomorphism of order three of the embedding illustrated in Figure 6.14. The homeomorphism that induces this automorphism rotates the graph by 120° about an axis perpendicular to the page, and containing the vertex 7. This example could easily be modified to increase the number of vertices that have an order of three or to add another fixed point. However, increasing the number of vertices complicates the illustration.

The main observation to make about Theorem 6.12 is that most automorphisms of complete graphs cannot be induced by a homeomorphism, no matter how the graph is embedded in $\mathbb{R}^3$. In particular, most automorphisms of $K_n$ whose orbits do not all have the same order, and any automorphism of $K_n$ that fixes more than three vertices, cannot be realized by a homeomorphism of $\mathbb{R}^3$, no matter how the complete graph is embedded. For example, the automorphism (12345)(678) of $K_8$ has an orbit of order 5 and an orbit of order 3, so this automorphism cannot be induced by an homeomorphism of $\mathbb{R}^3$. By applying an automorphism that cannot be induced by a homeomorphism of $\mathbb{R}^3$ to any embedding of a labeled complete graph, we can obtain a topological stereoisomer of the labeled graph. Furthermore, for any embedding of a complete graph on $4n + 1$ or $4n + 2$ vertices, an automorphism of order four can only satisfy condition 5 of Theorem 6.12, and thus can never be induced by an orientation-preserving homeomorphism. For $K_{4n+3}$, an automorphism of order four will not satisfy any of the conditions of Theorem 6.12. So, applying an order-four automorphism to any labeled embedding of $K_{4n+1}$, $K_{4n+2}$, or $K_{4n+3}$ will give us a topological stereoisomer.

We can use Theorem 6.12, together with the fact that the sum of the orders of all of the nontrivial orbits plus the number of fixed vertices must equal $n$, in order to enumerate all of the types of automorphisms that can be induced by a homeomorphism of $\mathbb{R}^3$ for any given $K_n$. For example, we obtain the following complete list of types of automorphisms, $\Phi$, of $K_7$ that can be induced by homeomorphisms of $\mathbb{R}^3$.

1. $\Phi$ has three orbits of order 2 and one fixed vertex.
2. $\Phi$ has two orbits of order 3 and one fixed vertex.
3. $\Phi$ has one orbit of order 5 and two fixed vertices.
4. $\Phi$ has one orbit of order 7 and no fixed vertices.

Thus $K_7$ has only four types of automorphisms that can be induced by homeomorphisms of $\mathbb{R}^3$, no matter how $K_7$ is embedded in $\mathbb{R}^3$.

From Theorem 6.1, we see that for $n > 6$, any automorphism of $K_n$ that is induced by a homeomorphism of $\mathbb{R}^3$ can actually be induced by a finite-order homeomorphism of $S^3$. Thus every automorphism satisfying one of the conditions listed in Theorem 6.12 is induced by a finite-order homeomorphism of $S^3$. It is natural to wonder whether each of these automorphisms can be induced by a finite-order homeomorphism of $\mathbb{R}^3$; or whether some of these automorphisms can be induced by a homeomorphism of some embedding of the graph in $\mathbb{R}^3$, but cannot be induced by a finite-order homeomorphism of $\mathbb{R}^3$, no matter how the graph is embedded. In answer to this question, we will prove the following theorem, which classifies those automorphisms that are induced by a finite-order homeomorphism of $\mathbb{R}^3$. The constructions that we give of the embeddings of the graphs in the proof of Theorem 6.13 (Flapan, 1995) will give the reader a sense of the types of constructions that we omitted from the proof of Theorem 6.12. The types of constructions that are not necessary for the proof of Theorem 6.13 yet are necessary for the proof of Theorem 6.12, are those involving finite-order orientation-preserving homeomorphisms of $S^3$ that have no fixed points. Such homeomorphisms do not induce a finite-order homeomorphism of $\mathbb{R}^3$.

**Theorem 6.13.** *Let* $K_n$ *be a complete graph on* $n > 6$ *vertices, and let* $\Phi$ *be an automorphism of* $K_n$ *of order* $m$, *for some* $m > 1$. *Then there is an embedding of* $K_n$ *in* $\mathbb{R}^3$ *such that* $\Phi$ *is induced by an orientation-preserving homeomorphism*

$h:(\mathbb{R}^3, K_n) \to (\mathbb{R}^3, K_n)$ *of finite order if and only if* $\Phi$ *satisfies the following condition:* (1) *either* m $= 2$ *or* m *is odd, and every nontrivial orbit of* $\Phi$ *has an order of* m, *and there are at most two fixed vertices. Furthermore, there is an embedding of* $K_n$ *in* $\mathbb{R}^3$ *such that* $\Phi$ *is induced by an orientation-reversing homeomorphism* $h:(\mathbb{R}^3, K_n) \to (\mathbb{R}^3, K_n)$ *of finite order if and only if* $\Phi$ *satisfies the following condition:* (2) m $= 4$, *either every nontrivial orbit of* $\Phi$ *has an order of four and* $\Phi$ *fixes at most one vertex, or every orbit of* $\Phi$ *has an order of four except one, which has an order of two.*

*Proof.* First we suppose that $\Phi$ is induced by a finite-order homeomorphism $h:(\mathbb{R}^3, K_n) \to (\mathbb{R}^3, K_n)$ for some embedding of $K_n$ in $\mathbb{R}^3$. We know that $\Phi$ must satisfy one of the conditions of Theorem 6.12. If $h$ is orientation reversing, $\Phi$ must satisfy condition 5 of Theorem 6.12, which is identical to condition 2 of Theorem 6.13. So in this case we are done.

Assume now that $h$ is orientation preserving. Then $\Phi$ satisfies one of conditions 1, 2, 3 or 4 of Theorem 6.12. Extend $h$ to a finite-order homeomorphism of $S^3$ by defining $h(\infty) = \infty$. First suppose that $m$ is even. Because $h$ is orientation preserving, $h$ has at least one fixed point (the point $\infty$), and since $m$ is even, it follows from Lemma 6.9 that $m = 2$. Thus $\Phi$ must satisfy condition 2 of Theorem 6.12. So $\Phi$ satisfies condition 1 of Theorem 6.13.

Now suppose that $m$ is odd. Then $\Phi$ satisfies either condition 3 or 4 of Theorem 6.12. Suppose that $\Phi$ satisfies condition 4; then $m > 3$ is an odd multiple of three and $\Phi$ has an orbit of order 3. So $\Phi^3$ fixes three vertices. Hence $\Phi^3$ also pointwise fixes the edges between these vertices. Now $h^3$ is a finite-order orientation-preserving homeomorphism of $S^3$ that pointwise fixes the simple closed curve consisting of these three vertices and the edges between them. However, $h^3$ must also fix the point $\infty$. We know that $\infty$ is not contained in the graph $K_n$ because $K_n$ is embedded in $\mathbb{R}^3$. Thus $h^3$ is an orientation-preserving finite-order homeomorphism that pointwise fixes a simple closed curve and a point not on the simple closed curve. This contradicts Smith's Theorem 4.1 unless $h^3$ is the identity. Hence $\Phi^3$ is the identity automorphism, and so $m = 3$, contradicting our assumption that $m > 3$.

Thus for any $m$ that is odd, $\Phi$ must satisfy condition 3 of Theorem 6.12. Now suppose that $\Phi$ fixes three vertices of $K_n$. Then, as above, $\Phi$ pointwise fixes the edges between these vertices, which implies that $h$ pointwise fixes a simple closed curve together with the point $\infty$. This again contradicts Smith's Theorem 4.1, because we are assuming that $\Phi$ is not the identity automorphism. Hence, $\Phi$ can fix at most two vertices. Therefore, $\Phi$ satisfies condition 1 of Theorem 6.13. So if $h$ has finite order and is orientation preserving, then $\Phi$ must satisfy condition 1 of Theorem 6.13, independent of whether $m$ is odd or even.

In order to prove the converse, we start with the assumption that $\Phi$ satisfies condition 1 or 2 of Theorem 6.13. If $\Phi$ satisfies condition 2, then $n$ cannot be of the form $4p + 3$ for any natural number $p$. Thus $n = 4p$, $n = 4p + 1$, or $n = 4p + 2$ for some natural number $p$. By examining the topologically achiral embeddings of $K_{4p}$, $K_{4p+1}$, and $K_{4p+2}$ that were constructed in Chapter 5, we see that for each of these complete graphs, the automorphism described by condition 2 of Theorem 6.13 is induced by an orientation-reversing finite-order homeomorphism of $\mathbb{R}^3$.

Now suppose that $\Phi$ satisfies condition 1 of Theorem 6.13. The embeddings of $K_n$ that we will construct will be somewhat similar to the topologically achiral embeddings of $K_{4n}$, $K_{4n+1}$, and $K_{4n+2}$, which we constructed in Chapter 5. We consider the cases in which $m = 2$ and in which $m$ is odd separately. First suppose that $m = 2$. We want to embed a complete graph $K_n$ in $\mathbb{R}^3$ in such a way that an automorphism $\Phi$ of order 2 with at most two fixed vertices is induced by a finite-order homeomorphism of $\mathbb{R}^3$. We can write $n = 2r$, $n = 2r + 1$, or $n = 2r + 2$ for some $r \in \mathbb{N}$, depending on whether $\Phi$ has zero, one, or two fixed vertices, respectively. We shall first explain how to embed $K_{2r+2}$ so as to induce an order-two automorphism with two fixed points by a finite-order homeomorphism.

Let $P$ be the $xy$ plane in $\mathbb{R}^3$ and let $C$ be the unit circle in $P$ centered at the origin. Also, let $h : \mathbb{R}^3 \to \mathbb{R}^3$ be a 180° rotation about the $z$ axis. Pick consecutive vertices $v_1, \ldots, v_{2r}$ on $C$ in such a way that for each $i \leqslant 2r$ we have $h(v_i) = v_{i+r}$, where the subscripts are considered mod $2r$. Pick vertices $v_{2r+1}$ and $v_{2r+2}$ to be two distinct points on the $z$ axis. Let the edge $\overline{v_{2r+1}v_{2r+2}}$ (joining vertices $v_{2r+1}$ and $v_{2r+2}$) be the line segment on the $z$ axis joining $v_{2r+1}$ and $v_{2r+2}$. For every $i \leqslant 2r$, let the edge $\overline{v_i v_{2r+1}}$ be the straight line segment joining $v_i$ to $v_{2r+1}$, and let the edge $\overline{v_i v_{2r+2}}$ be the straight line segment joining $v_i$ to $v_{2r+2}$. Now for each $i \leqslant r$, pick points $x_i$ on the $z$ axis that are disjoint from the edge $\overline{v_{2r+1}v_{2r+2}}$. For each $i \leqslant r$, let $d_i$ denote the straight line segment joining vertex $v_i$ to the point $x_i$. Now $e_i = d_i \cup h(d_i)$ will be a path joining the vertex $v_i$ and the vertex $v_{i+r}$, and $h(e_i) = e_i$. For each $i \leqslant r$, let the edge $\overline{v_i v_{i+r}}$ be the path $e_i$. Observe that for every $i \leqslant 2r$ and $j \leqslant 2r$, the distance between vertices $i$ and $j$ on $C$ is no more than $r$ edges if we measure the distance around $C$ in the shortest direction.

Now choose nested ellipsoids $E_1, \ldots, E_r$, which all contain the circle $C$ but are otherwise disjoint, are each setwise invariant under $h$, and do not intersect the interiors of the edges that we have already chosen except those contained in $C$. Also for each $i = 1, \ldots, r$ and each $j$ such that $i < j < i + r$, let $F_{ij}$ denote a plane that contains vertices $i$ and $j$ and is perpendicular to the plane $P$. We construct the edge $f_{ij}$ connecting vertex $i$ to vertex $j$ to be the arc in the intersection of $E_i$ and $F_{ij}$ on the positive side of the $xy$ plane. Note that for a fixed $i$ and any $j$ such that $i < j < i + r$, the edges $f_{ij}$ only intersect at the vertex $i$. Also if $i \neq i'$, then $f_{ij}$ is contained in $E_i$, and $f_{i'j'}$ is contained in $E_{i'}$. Thus the edges $f_{ij}$ and $f_{i'j'}$ are disjoint, except possibly at the vertex $j$ if $j = j'$. Recall that $h$ rotates $C$ by $180°$, leaving each $E_i$ setwise invariant. Hence for a fixed $i \leqslant r$, and for every $j$ and $k$ such that $i < j < i + r$ and $i < k < i + r$, the image $h(f_{ij})$ will be contained in $E_i$ and be disjoint from $f_{ik}$. We now add to our construction the image of each of the edges $f_{ij}$ under $h$, as the edge that joins vertices $i + r$ and $j + r$. In this way we obtain edges between every pair of vertices (readers should convince themselves of this). By our construction, the set of edges will all be disjoint, and this embedding of $K_{2r+2}$ will be setwise invariant under $h$. Also $h$ will fix each of the vertices $v_{2r+1}$ and $v_{2r+2}$ and will fix no other vertices of $K_{2r+2}$.

Suppose that we want to embed $K_{2r+1}$ in such a way that an automorphism of order 2 with only one vertex fixed is induced by an order-two homeomorphism of $\mathbb{R}^3$. Then we start with the above embedding of $K_{2r+2}$ and the above homeomorphism $h$, and we omit the vertex $v_{2r+2}$ together with all the edges containing it. To get an embedding of $K_{2r}$ such that the automorphism of order two with no fixed vertices is induced by a finite-order homeomorphism of $\mathbb{R}^3$, we omit both vertices $v_{2r+2}$ and $v_{2r+1}$ and the edges containing them from the above embedding of $K_{2r+2}$. So if $\Phi$ has an order of two and satisfies condition 1 of Theorem 6.13, then $\Phi$ can be induced by an order-two homeomorphism of $R^3$, for the embedding of $K_n$ that we have given above.

Now suppose that $\Phi$ has an odd order of $m$ and satisfies condition 1 of Theorem 6.13. We begin as we did above, but then the construction will differ significantly. We first assume that $\Phi$ has two fixed vertices, so $n = mr + 2$ for some $r \in \mathbb{N}$.

Let $P$ be the $xy$ plane in $\mathbb{R}^3$ and let $C$ be the unit circle in $P$ centered at the origin. Also, let $h : \mathbb{R}^3 \to \mathbb{R}^3$ be a $(2\pi/m)$ rotation about the $z$ axis. Thus $h$ has an order of $m$. Pick $m$ disjoint arcs $a_1, \ldots, a_m$ on $C$, each of length $\pi/m$, with spaces of length $\pi/m$ separating each $a_i$ from the proceeding one. Then $h(a_k) = a_{k+1}$ for each $k \leqslant m$, where the subscripts are considered mod $m$. Now within the arc $a_1$ pick vertices $v_1, \ldots, v_r$, and for each $i \leqslant r$ and $k < m$, let $v_{i+kr} = h^k(v_i)$. Pick vertices $v_{mr+1}$ and $v_{mr+2}$ to be two disjoint points on the $z$ axis. Let the edge $\overline{v_{mr+1}v_{mr+2}}$ be the line segment on the $z$ axis joining $v_{mr+1}$ and $v_{mr+2}$. For every $i \leqslant mr$, let the edge $\overline{v_i v_{mr+1}}$ be the straight line segment joining $v_i$ to $v_{mr+1}$, and let the edge $\overline{v_i v_{mr+2}}$ be the straight line segment joining $v_i$ to $v_{mr+2}$.

For each $i \leqslant r$ and $j$ such that $i < j \leqslant i + mr/2$, let $E_{ij}$ denote an ellipsoid containing $C$, which is disjoint from $\overline{v_{mr+1}v_{mr+2}}$ and which is invariant under

$h$. Pick the collection of ellipsoids $E_{ij}$ so that they are pairwise disjoint except along $C$. For each $i \leqslant r$ and $j$ such that $i < j \leqslant i + mr/2$, let $F_{ij}$ denote the plane containing vertices $v_i$ and $v_j$ that is perpendicular to $P$. Now let $f_{ij}$ denote the intersection of the plane $F_{ij}$ with the ellipsoid $E_{ij}$. Observe that because the $E_{ij}$ are all disjoint except along $C$, for any $i \leqslant r$ and $j$ such that $i < j \leqslant i + mr/2$ and any $i' \leqslant r$ and $j'$ such that $i' < j' \leqslant i' + mr/2$, we know that $f_{ij}$ and $f_{i'j'}$ are disjoint if $i \neq i'$ or $j \neq j'$. The arc $f_{ij}$ will not be one of our edges but will help us build our edges. For each $k < m$, let $f_{ijk}$ denote $h^k(f_{ij})$. Each $f_{ijk}$ is an arc in $E_{ij}$ with endpoints at vertices $v_{i+kr}$ and $v_{j+kr}$. It can be shown that the arc $f_{ij}$ may intersect $f_{ijk}$; however, for a fixed $i$ and $j$, if $k \neq l$, then $f_{ijk}$ and $f_{ijl}$ do not both intersect $f_{ij}$ at the same point (the reader will prove this assertion in Exercise 11). Observe that because $m$ is odd, for each arc $a_i$ on $C$ there is no arc $a_j$ opposite it on $C$. So, no pair of vertices $v_i$ and $v_j$ are antipodal points on $C$. It follows that no $f_{ij}$ or $f_{ijk}$ intersects the $z$ axis. Thus no point of $f_{ij}$ is fixed by any power of $h$.

Now we alter $f_{ij}$ near one point where some $f_{ijl}$ intersects it by replacing a small piece of $f_{ij}$ by a small arc in the plane $F_{ij}$, going slightly above the ellipsoid $E_{ij}$, but not so much above it that it intersects another one of our ellipsoids. In Figure 6.15 we illustrate how the small piece of $f_{ij}$ is replaced. By an abuse of notation, we also call this new arc $f_{ij}$, and for each $k < m$ the image $h^k$ of the new $f_{ij}$ will also be called $f_{ijk}$. By modifying $f_{ij}$ and all of the $f_{ijk}$ in this way, we eliminate a point where $f_{ij}$ intersected $f_{ijl}$ and another point where $f_{ij}$ intersected $f_{ij(m-l)}$. By making our modification small enough we do not introduce any new points of intersection. Now if there exists any point where our new $f_{ij}$ intersects one of our new $f_{ijk}$, we again alter $f_{ij}$ as above. In this way we again obtain a new arc $f_{ij}$ and a collection of new arcs $f_{ijk}$. Each time we do this alteration of $f_{ij}$, we reduce the total number of points of intersection of $f_{ij}$ with any $f_{ijk}$. So, eventually, $f_{ij}$ will become disjoint from $f_{ijk}$ for every $k < m$. Let this final $f_{ij}$ be the edge in $K_n$ from $v_i$ to $v_j$, and for every $k < m$ let $f_{ijk}$ be the edge $h^k(f_{ij})$ from $v_{i+kr}$ to $v_{j+kr}$. Observe that because $f_{ij}$ is disjoint from every $f_{ijk}$, any pair $f_{ijk}$ and $f_{ijl}$ will also be disjoint from each other.

In this way, we have constructed disjoint edges between every pair of vertices $v_i$ and $v_j$, thus completing our embedding of $K_{mr+2}$ (the reader should check this).

Suppose that we want to embed $K_{mr+1}$ in such a way that an automorphism with an order of $m$ with only one vertex fixed is induced by an order-$m$ homeomorphism of $\mathbb{R}^3$. Then we start with the above embedding of $K_{mr+2}$ and the above homeomorphism $h$, and we omit the vertex $v_{mr+2}$ together with all the edges containing it. To get an embedding of $K_{mr}$ such that an automorphism of

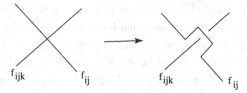

**Figure 6.15.** Wherever $f_{ij}$ intersects $f_{ijk}$, we replace a small arc of $f_{ij}$ by an arc passing above $f_{ijk}$

order $m$ with no fixed vertices is induced by an order-$m$ homeomorphism of $\mathbb{R}^3$, we omit both vertices $v_{mr+2}$ and $v_{mr+1}$ and the edges containing them from the above embedding of $K_{mr+2}$. So if $\Phi$ has odd order $m$ and satisfies condition 1 of Theorem 6.13, then $\Phi$ can be induced by an order-$m$ homeomorphism of $\mathbb{R}^3$, for the embedding of $K_n$ that we have given above.                                    □

By comparing Theorem 6.13 with Theorem 6.12, we see that there exist automorphisms of a complete graph $K_n$ that can be induced by a homeomorphism of $\mathbb{R}^3$, for some embedding of $K_n$ in $\mathbb{R}^3$, but cannot be induced by a finite-order homeomorphism of $\mathbb{R}^3$, no matter how $K_n$ is embedded. For example, let $\Phi$ be the automorphism of $K_8$ with one orbit of order 5 and three fixed vertices. By Theorem 6.13, this automorphism cannot be induced by a finite-order homeomorphism of $\mathbb{R}^3$ no matter how $K_8$ is embedded. In contrast, Figure 6.16 illustrates an embedding of $K_8$ in $\mathbb{R}^3$ such that there is a homeomorphism $h: (\mathbb{R}^3, K_8) \rightarrow (\mathbb{R}^3, K_8)$ that induces the automorphism (12345). The homeomorphism $h$ can be seen in Figure 6.17 as a rotation by 72° about the central axis containing vertices 6, 7, and 8, followed by a deformation of the edge $\overline{78}$ that takes it back to its original spot. Because of this last deformation, this homeomorphism of $\mathbb{R}^3$ is not of finite order.

### Hierarchies of Automorphisms

We have seen examples of automorphisms that cannot be induced by a homeomorphism of $S^3$, no matter how the graph is embedded in $S^3$, as well as automorphisms that can be induced by a homeomorphism of $S^3$ for some embedding of the graph in $S^3$, but cannot be induced by a finite-order homeomorphism of $S^3$ no matter how the graph is embedded in $S^3$. We can rank automorphisms according to how they can be induced by a homeomorphism in $S^3$ or in $\mathbb{R}^3$. In particular, the following hierarchy of automorphisms can be considered in $S^3$ or in $\mathbb{R}^3$ :  (1) automorphisms that cannot be induced by a

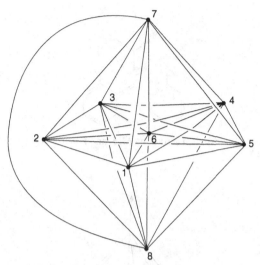

**Figure 6.16.** The automorphism (12345) can be induced by a homeomorphism of $\mathbb{R}^3$ but not by a finite-order homeomorphism of $\mathbb{R}^3$

homeomorphism, no matter how the graph is embedded; (2) automorphisms that can be induced by a homeomorphism for some embedding, but cannot be induced by a finite-order homeomorphism, no matter how the graph is embedded; and (3) automorphisms that can be induced by a finite-order homeomorphism for some embedding of the graph.

Those automorphisms that are not listed in Theorem 6.12 give us many examples of automorphisms of complete graphs that are in category 1 of this hierarchy when considered in either $S^3$ or $\mathbb{R}^3$. Because an automorphism of a graph can be induced by a homeomorphism of some embedding of the graph in $S^3$ if and only if it can be induced by a homeomorphism of some embedding of the graph in $\mathbb{R}^3$, the class of automorphisms in category 1 of the hierarchy is the same whether we are embedding our graphs in $S^3$ or in $\mathbb{R}^3$.

We saw an example at the beginning of the chapter (illustrated again in Figure 6.17) of an automorphism of a labeled graph that can be induced by a homeomorphism but not by a finite-order homeomorphism of $S^3$ or $\mathbb{R}^3$. This automorphism is of category 2 of the hierarchy when considered in either $S^3$ or $\mathbb{R}^3$. Because any finite-order homeomorphism $h$ of $\mathbb{R}^3$ can be extended to a finite-order homeomorphism of $S^3$ by defining $h(\infty) = \infty$, every automorphism of category 2 in the hierarchy for graphs that are embedded in $S^3$ is necessarily of category 2 of the hierarchy for graphs that are embedded in $\mathbb{R}^3$.

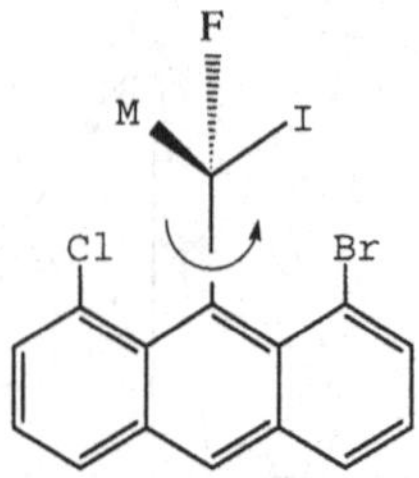

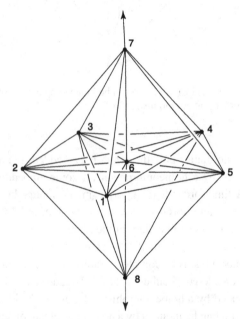

**Figure 6.17.** The automorphism $\varphi$ rotating M, I, and F is in category 2 of the hierarchy in either $S^3$ or $\mathbb{R}^3$

**Figure 6.18.** The automorphism (12345) of $K_8$ can be induced by a rotation if the graph is embedded in $S^3$

By our comments at the end of the last section, the automorphism (12345) of $K_8$ is in category 2 of our hierarchy for graphs embedded in $\mathbb{R}^3$ (see Figure 6.16). However, all of the automorphisms listed in Theorem 6.12 can be induced by a finite-order homeomorphism of $S^3$ for some embedding of $K_n$ in $S^3$. In particular, the automorphism (12345) of $K_8$ can be induced by a finite-order homeomorphism of $S^3$ if $K_8$ is embedded as illustrated in Figure 6.18. The embedding is the same as that of Figure 6.16 except that we chose the point at infinity to be the midpoint of the edge $\overline{78}$. Now a rotation by

72° about the central axis will induce (12345). All of the automorphisms that are listed in Theorem 6.12 but not in Theorem 6.13 are in category 2 of the hierarchy in $\mathbb{R}^3$, but in category 3 of the hierarchy in $S^3$. So, automorphisms in category 2 in $S^3$ are a proper subset of automorphisms in category 2 of $\mathbb{R}^3$.

There are many examples of automorphisms in category 3 in $\mathbb{R}^3$. For example, any automorphism that can be induced by a rotation or a reflection is in category 3. Any automorphism that is in category 3 in $\mathbb{R}^3$ is necessarily in category 3 in $S^3$, but not vice versa. Thus, automorphisms in category 3 in $\mathbb{R}^3$ are a proper subset of automorphisms in category 3 in $S^3$.

Suppose that $\Phi$ is an automorphism of a labeled graph $G$ such that $\Phi$ can be induced by an orientation-preserving homeomorphism for some embedding of $G$, but not by a finite-order homeomorphism for this embedding. Then performing $\Phi$ on this embedding of $G$ will create a topologically rigid stereoisomer of $G$ that is not a topological stereoisomer of $G$. The existence of many automorphisms that are in category 2 of the hierarchy when considered in $\mathbb{R}^3$ allows us to easily create many examples of this type in $\mathbb{R}^3$. In order to create such examples in $S^3$ we must either use graphs that are not three connected or else choose the embedding of the graph, as we did in Figure 6.7, so that the particular embedding has an automorphism that is induced by a homeomorphism but not by one that is of finite order.

## Exercises

1. Prove that the automorphism (12)(345) of the following graph can be induced by a homeomorphism for the given embedding, but cannot be induced by a finite-order homeomorphism for any embedding.

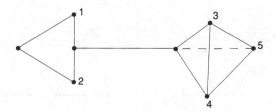

2. Prove that any Möbius ladder with two or more rungs is three connected.
3. Prove that any complete graph with four or more vertices is three connected.
4. Explain why every automorphism of $K_5$ can be induced by a homeomorphism of the embedding that is illustrated in Figure 6.12.
5. Prove that every automorphism of $K_6$ except those conjugate to (1234) can be induced by a homeomorphism for some embedding of $K_6$ in $S^3$.
6. Prove that the molecular graph of ferrocenophan has no nontrivial automorphisms.

7. Suppose that we remove all of the labels from the vertices of the molecular graph of ferrocenophan. Would this unlabeled graph have an automorphism of order two? Would it still be intrinsically chiral? Explain.

8. Suppose that we remove all of the labels from the vertices of the graph of the Simmons–Paquette molecule. Would this unlabeled graph have an automorphism of order two? Would it still be intrinsically chiral? Explain.

9. Prove that the protein molecule that is illustrated in Figure 6.11 is intrinsically chiral.

10. Suppose that $f$ is a finite-order homeomorphism of an interval that fixes each of the endpoints. Prove that $f$ is the identity map.

11. Let $h$ be a rotation of $\mathbb{R}^3$ about the $z$ axis around an angle of $(2\pi/m)$ for some odd number $m$. Let $P$ be the $xy$ plane in $\mathbb{R}^3$ and let $C$ be the unit circle in $P$ centered at the origin. Pick $m$ disjoint arcs $a_1, \ldots, a_m$ on $C$, each of length $\pi/m$, such that $h(a_k) = a_{k+1}$ for each $k \leqslant m$, where the subscripts are considered mod $m$. Now within the arc $a_1$ pick vertices $v_1, \ldots, v_r$, and for each $i \leqslant r$ and $k < m$, let $v_{i+kr} = h^k(v_i)$. Consider a straight line segment $e_1$, in the $xy$ plane with vertices $v_i$ and $v_j$, where $i \leqslant r$ and $j \leqslant i + mr/2$. Let $h^k(e_1) = e_2$ and $h^l(e_1) = e_3$ for some $l$ and some $k \neq l$. We wish to prove that $e_1 \cap e_2 \cap e_3$ must be empty. Suppose that all three of these edges meet in a point $x$: (a) prove that if the vertices of the three edges occur in the order given in any of the three figures below, then $x$ must be at the center of the circle; (b) prove that $e_1 \cap e_2 \cap e_3$ must, in fact, be empty; and (c) use part (b) to prove that in the construction given in the proof of Theorem 6.13 for $m$ odd, the arcs $f_{ijk}$ and $f_{ijl}$ do not both intersect $f_{ij}$ at the same point.

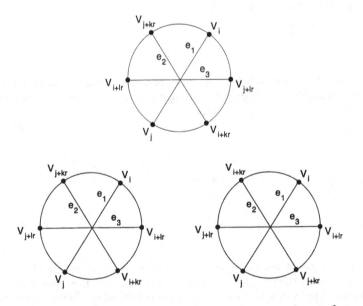

12. Find an embedding of $K_6$ such that there is a homeomorphism $h : (S^3, K_6) \rightarrow (S^3, K_6)$ of order 6, that has no fixed points but such that $h^2$ does have fixed points.

13. Give a complete list of the types of automorphisms of $K_8$ and of $K_9$ that can be induced by a homeomorphism of $\mathbb{R}^3$ for some embedding of $K_8$ and of $K_9$ in $\mathbb{R}^3$.

14. Find an embedding of $K_7$ in $\mathbb{R}^3$ such that the automorphism (1234567) is induced by a homeomorphism of $\mathbb{R}^3$.

15. Find an embedding of $K_8$ in $S^3$ such that the automorphism (1234)(5678) is induced by an orientation-preserving homeomorphism of $S^3$, and find another embedding of $K_8$ in $S^3$ such that the automorphism (1234)(5678) is induced by an orientation-reversing homeomorphism of $S^3$.

16. Prove that no three-connected graphs can be in category 2 of our hierarchy of achiral embeddability in $S^3$ (see Chapter 4 for an explanation of this hierarchy), and no automorphism of a three-connected graph can be in category 2 of our hierarchy of automorphisms in $S^3$.

17. Suppose that $B$ is a three-dimensional ball, and $f : \partial B \to \partial B$ is a finite-order homeomorphism. Prove that $f$ can be extended to a finite-order homeomorphism of $B$.

18. Suppose that $V$ is a solid torus, and $f : \partial V \to \partial V$ is a finite-order homeomorphism that takes meridians to meridians. Prove that $f$ can be extended to a finite-order homeomorphism of $V$.

19. Suppose that $V$ is a solid torus that is fibered of type $(p, q)$ and $W$ is a solid torus that is fibered of type $(p'q')$. A homeomorphism $h : V \to W$ is said to be *fiber preserving* if, for every fiber $f_x$ of $V$, we have that $h(f_x)$ is a fiber of $W$. Prove that there exists a fiber-preserving homeomorphism $h : V \to W$ if and only if $p = p'$ and $q = q'$.

# 7

# Topology of DNA

Topological techniques are well suited to the study of DNA, because DNA molecules are both long and flexible. The basic structure of duplex DNA consists of two molecular strands that are twisted together in a right-handed helix, while the two strands are joined together by bonds. Figure 7.1 illustrates how the basic structure of duplex DNA resembles a twisted ribbon.

Each strand of duplex DNA is made up of alternating sugars and phosphates, and every sugar has one of four bases attached to it. The bases (illustrated in Figure 7.2) are called *adenine, thymine, cytosine,* and *guanine,* and they are normally designated only by the first letters of their names. The letter R on each base indicates where the base is attached to a sugar.

There are bonds that attach each base on one of the strands to a base on the other strand, and thus hold the two strands together. These bonds are formed because one of the hydrogen atoms of an NH or an $NH_2$ on one base bonds with an N or an O, respectively, on the other base. The base A on one strand is always bonded with the base T on the other strand, and the base C on one strand is always bonded with the base G on the other strand. In Figure 7.3 we illustrate the hydrogen bonding between the bases A and T, and between the bases C and G. A pair of bases, which are bonded together, is referred to as a *base pair* of the molecule. Observe that because of the way the bases are bonded, the sequence of bases that occurs on a single strand determines the sequence that occurs on the other strand. So we can speak of the sequence of base pairs, or simply the genetic sequence, rather than having to specify the sequence on each strand.

The strands of DNA are each made up of small units called *nucleotides,* consisting of one sugar, one phosphate, and one base. Figure 7.4 illustrates the sugar and phosphate molecules contained in the nucleotide. Each sugar molecule has a particular site on it that is referred to as the 3′ site and another site called the 5′ site. Within a strand of DNA, each phosphate is joined to the 3′ site of one sugar and to the 5′ site of another sugar. One end of a strand of

**Figure 7.1.** The basic structure of DNA is that of a twisted ribbon

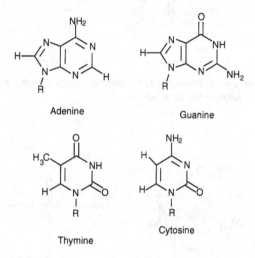

Adenine

Guanine

Thymine

Cytosine

**Figure 7.2.** The bases cytosine, thymine, adenine, and guanine

DNA contains the 3′ site of a sugar and is called the 3′ end, while the other end of the strand contains the 5′ site of another sugar and is called the 5′ end of the strand. The two strands of duplex DNA are oppositely oriented, so that one end of the duplex consists of the 5′ end of one strand and the 3′ end of the other strand, while the other end of the DNA consists of the 3′ end of the first strand and the 5′ end of the second strand. Figure 7.5 illustrates two nucleotides on a single strand, with the 5′ end and the 3′ end marked.

Duplex DNA can exist in nature in linear form (i.e., as a long line segment) or in closed circular form (i.e., as a simple closed curve), although the linear form is far more common. To give some idea of the length of DNA, the human genome contains about $3 \times 10^9$ base pairs. Because a phosphate cannot bond simultaneously to two 3′ sites, a single strand of closed circular DNA must be

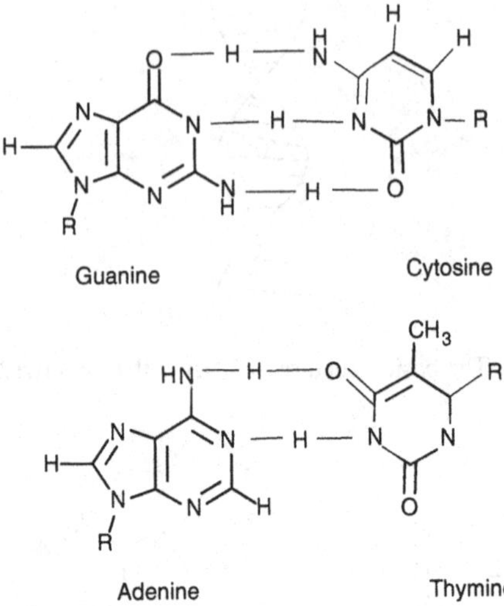

Guanine                    Cytosine

Adenine                    Thymine

**Figure 7.3.** There are hydrogen bonds between the bases guanine and cytosine and between the bases adenine and thymine

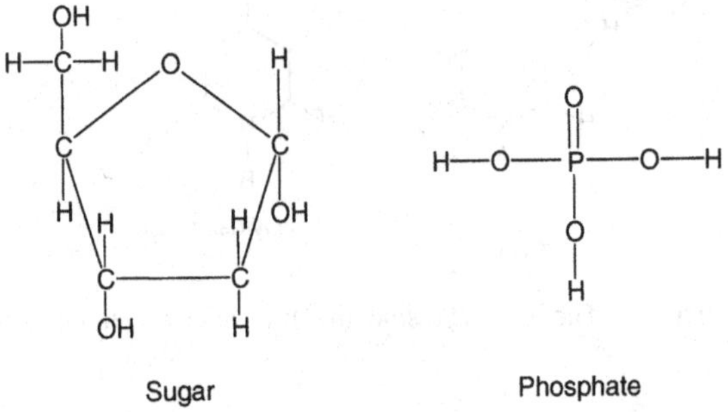

Sugar                    Phosphate

**Figure 7.4.** A sugar and a phosphate

joined to itself rather than to the other strand of the duplex DNA. Hence a closed circular DNA molecule always has the form of a twisted annulus rather than of a twisted Möbius band. Such a DNA molecule can be knotted, and two or more circular DNA molecules can be linked together. Both knotted and linked duplex DNA can be found in nature, though links are much more common than knots. In fact, linked duplex DNA molecules are known to occur naturally in human cells. In addition to considering the topology of two DNA molecules that are linked, we can consider the linking of the two strands within a single closed circular duplex DNA molecule. If the closed circular DNA is not knotted, then

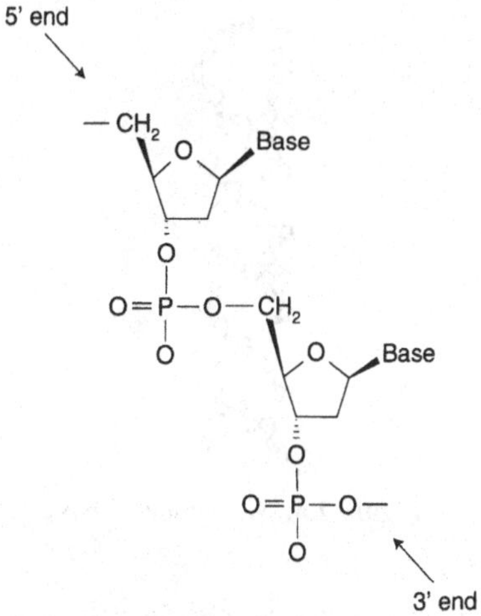

5' end

$$-CH_2 \quad O \quad Base$$

O

O=P—O—$CH_2$
|
O  O  Base

O

O=P—O—
|
O

3' end

**Figure 7.5.** A chain of two nucleotides

the two strands, which are twisted together, have the form of a $(2, n)$-torus link in which $n$ is quite large (see Chapter 2 for the definition of a torus link). Because, for any $n > 1$, a $(2, n)$-torus link is topologically chiral, it follows that all closed circular duplex DNA molecules are topologically chiral, and hence chemically chiral. The linking number between the two strands can help distinguish DNA molecules. In particular, if two DNA molecules have the same sequence of base pairs but the pairs of strands have different linking numbers, then the molecules are topological stereoisomers, and hence chemical stereoisomers.

## Supercoiling

If we think of duplex DNA as a ribbon, then the central line of the ribbon is referred to as the *helix axis* of the DNA molecule, and the two strands that make up the sides of the ribbon are called the *backbone* of the DNA. A DNA molecule is said to be in a *relaxed state* if it twists in a right-handed helix with approximately one full turn for every 10.5 base pairs. In this state, the helix axis is roughly a planar circle, though the strands form a torus link and hence they do not lie in a plane. If more or fewer twists are introduced between the strands, then in order to minimize the torsional stress, the DNA will become supercoiled (i.e., the helix axis becomes coiled up like an overused telephone cord). Figure 7.6 gives a rough idea of what happens when a pair of strands are supercoiled. Another way in which DNA can become supercoiled is by wrapping itself around a protein molecule. When DNA is supercoiled, it takes up less space than relaxed DNA and hence is more efficiently stored in vivo. Almost all DNA molecules

**Figure 7.6.** Supercoiled DNA

are supercoiled at least part of the time. Supercoiling has an impact on most biological, chemical, and physical properties of DNA. In particular, it facilitates recombination by bringing together distant sites on the DNA molecule. There are many other important consequences of supercoiling, including the ways in which proteins bind to DNA. The mathematics of supercoiling involves both topology and differential geometry, as we will explain below. Because we are focusing on topology in this book rather than on differential geometry, we will not provide all of the details of the differential geometry. For more information about the biology and geometry of supercoiling than will be given in this chapter, see Bates and Maxwell (1993) and White (1995).

A basic tool for understanding supercoiling is the concept of the linking number. We defined this concept originally in Chapter 2; however, we restate the definition below.

**Definition.** Let $K_1$ and $K_2$ be oriented knot projections. We define the *linking number* of $K_1$ and $K_2$, written $\text{Lk}(K_1, K_2)$, to be one-half of the sum of $+1$ for every positive crossing between $K_1$ and $K_2$ and $-1$ for every negative crossing between $K_1$ and $K_2$.

We arbitrarily give the graph of a closed circular duplex DNA molecule an orientation, which in turn induces parallel orientations on the two strands and on the helix axis. The linking number of the DNA, which is denoted by Lk, actually is the linking number between the two oriented strands. Because the linking number is a topological invariant of a link, Lk will not change when the DNA deforms itself, as long as neither strand of the DNA is broken. Because both strands of the DNA are oriented in parallel directions, all of the crossings in a right-handed helix will be positive and all of the crossings in a left-handed helix will be negative. Hence, a right-handed helix will always have a positive linking number, whereas a left-handed helix will always have a negative linking number. Relaxed DNA has approximately 10.5 base pairs for every right-handed full turn, so a relaxed duplex DNA circle containing 1050 base pairs will have $Lk = 100$. If the number of base pairs in relaxed DNA is not evenly divisible by 10.5, then when the ends of the DNA are joined together, a small amount of twisting is necessarily added or subtracted so that the two ends of the DNA can be aligned. In general, if a relaxed closed circular DNA molecule contains $N$ base pairs, then its linking number will be the closest integer to $N/10.5$.

In 1969, James White proved that the linking number of DNA can be split into two quantities called the *average writhe* and the *twist*, which are associated with the geometry of the DNA (White, 1969). White's idea is that part of the linking between the two strands comes from the wrapping of the helix axis around itself, whereas another part comes from the twisting of the backbone of the DNA around the helix axis. In order to understand the concept of the average writhe of a DNA molecule, we first review the following definition from Chapter 2.

**Definition.** Consider a projection of an oriented link $L$. We define the *writhe* of this projection as the sum of $+1$ for every positive crossing of the link and $-1$ for every negative crossing of the link.

In order for the writhe to be well defined, the projection must be well defined in the sense that at most two points of the link lie over a single point of the plane. We saw in Chapter 2 that different projections of a knot may have different writhes. We use the notion of the writhe of a projection to define the average writhe of a DNA molecule.

**Definition.** The *average writhe* of a DNA molecule, denoted by Wr, is defined to be the average of the writhes of the helix axis over all possible well-defined projections of the helix axis.

Observe that if the helix axis is a planar circle, as it is for relaxed DNA, then $Wr = 0$, because no projection of it has any crossings except when viewed within the plane itself, which is not a well-defined projection. Although relaxed circular DNA is roughly a planar circle, it may not always be strictly planar, because it may contain some regions of local curvature. If this is the case, then the average writhe could be nonzero. The reader should note that, in general, unlike the linking number, which is always an integer, the average writhe of a DNA molecule is not necessarily an integer.

The goal of the twist is to measure how the backbone of the DNA wraps around the helix axis. We shall use the construction of White and Bauer (1986) to define the twist in terms of vectors. Let $f$ denote the helix axis and let $g$ denote the backbone of the DNA molecule. Before we begin, we should note that whereas $g$ is a strand of the DNA and hence is physically part of the DNA, $f$ is not a physical part of the molecule. Rather, $f$ comes from the way that we model the DNA as a ribbon. Thus there is some ambiguity about the definition of $f$. In particular, if a DNA molecule were modeled differently it might have a different $f$. However, we shall proceed as if the ribbon model of DNA is actually a physical reality.

We parametrize the helix axis $f$ by a variable $t \in [0, 1]$, and for each value of $t$, we let $P(t)$ denote the plane perpendicular to $f$ at the point $f(t)$. Note that $f(0) = f(1)$, and we assume that $f(t)$ is smooth even at the point $f(0)$. Now we parametrize $g$ by $t$ in such a way that for each $t$, the plane $P(t)$ intersects $g$ at $g(t)$. Then for each $t \in [0, 1]$, we define $v(t)$ to be the unit vector from $f(t)$ in the direction of $g(t)$. That is, $v(t) = g(t) - f(t)/ \|g(t) - f(t)\|$. This is well defined because $g(t)$ is never equal to $f(t)$. Also by definition $v(t)$ is always contained in $P(t)$. Because $v(t)$ is a unit vector, $v(t) \cdot v(t) = 1$. So taking the derivative of this dot product equation with respect to $t$, we have $2v(t) \cdot v'(t) = 0$. Let $v_1(t)$ denote the component of $v'(t)$ that is parallel to $f'(t)$, and let $v_2(t)$ denote the component of $v'(t)$ that is contained in the plane $P(t)$. Because $v(t)$ is contained in $P(t)$ and $f$ is perpendicular to $P(t)$, we know that $v_1(t)$ is perpendicular to $v(t)$. Hence we have both $v(t) \cdot v'(t) = 0$ and $v(t) \cdot v_1(t) = 0$. However, we know that $v'(t) = v_1(t) + v_2(t)$, so $v(t) \cdot v_2(t) = 0$. This means that $v_2(t)$ is perpendicular to $v(t)$ and lies in the plane $P(t)$.

We orient the two sides of the plane $P(t)$ so that $f(t)$ crosses $P(t)$ by going from the negative side of $P(t)$ to the positive side of $P(t)$. Let $w(t)$ denote the unit vector in $P(t)$ that forms an angle of $+\pi/2$ with $v(t)$, as seen from the positive side of $P(t)$. Then there is a scalar function $\lambda(t)$ such that $v_2(t) = \lambda(t)w(t)$. Intuitively, $\lambda(t)$ measures how much $v(t)$ is twisting around $f(t)$ at time $t$. Formally, we define the *twist* of the DNA to be $\mathrm{Tw} = 1/2\pi \int \lambda(t) \, dt$. To make this integral easier to compute, we observe that by taking the dot product of $v_2(t)$ with $w(t)$ we get $\lambda(t) = v_2(t) \cdot w(t) = [v_1(t) + v_2(t)] \cdot w(t) = v'(t) \cdot w(t)$. Thus $\mathrm{Tw} = 1/2\pi \int [v'(t) \cdot w(t)] \, dt$.

If the helix axis is a planar circle, then we can understand the twist as follows. Let $S$ denote the plane containing the helix axis and for each $t \in [0, 1]$ let $u(t)$ denote the unit vector in the normal vector field to $S$ at the point $f(t)$. Because $P(t)$ is the plane perpendicular to $f$ at the point $f(t)$, both $v(t)$ and $u(t)$ lie in

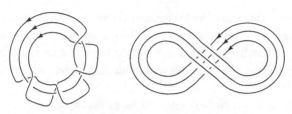

**Figure 7.7.** Examples of Lk, Wr, and Tw

$P(t)$. Let $\varphi(t)$ denote the angle measured from $u(t)$ to $v(t)$, as seen from the positive side of the plane $P(t)$. Now a computation shows that $\lambda(t) = \varphi'(t)$ for all $t$. Hence $Tw = 1/2\pi \int \varphi'(t)\,dt$. Observe that because the DNA is closed, $v(0) = v(1)$ and $u(0) = u(1)$, so the total change in $\varphi$ must be a multiple of $2\pi$. It follows that whenever the helix axis is a planar circle, the Tw will be an integer. In particular, when the helix axis is planar, the twist represents the number of times that the backbone wraps around the helix axis of the DNA. However, in general, when the helix axis is not planar, the twist will not be an integer.

In order to better understand the three quantities Lk, Wr, and Tw and their relationship with each other, consider the ribbons drawn in Figure 7.7. These examples are not meant to represent actual DNA molecules, rather they are meant to give the reader some intuition about the concepts we are discussing. In both examples, the helix axis is the central circle and the two strands are on either side of it. For the ribbon on the left, the axis lies in the plane of the paper, $Lk = 2$, $Wr = 0$, and $Tw = 2$. For the ribbon on the right, $Lk = -1$, $Wr \approx -1$ because most projections look like this one and have a writhe equal to $-1$, and $Tw \approx 0$ because the helix axis is almost planar and the backbone does not wrap around the helix axis. Observe that, although the axis of the ribbon on the right is topologically equivalent to a planar circle, the curve itself does not actually lie in a plane. Unlike in previous chapters, we are now considering the geometry of the structure as well as the topology, so we cannot just deform the axis so that it will lie in the plane.

White (1969) discovered a profound relationship among the concepts of linking number, average writhe, and twist, which is expressed by the equation $Lk = Tw + Wr$. Whereas Lk is easy to compute, in most cases, Tw is rather difficult to compute, and Wr is extremely difficult to compute. If we are able to compute the twist, then White's equation enables us to determine the average writhe.

Suppose that we start with relaxed linear DNA, and add or subtract twists to it, and then close it up. In order to minimize its torsional stress, the writhe of the molecule will change. In order to characterize how far a DNA molecule is from being relaxed, we want to compare the linking number of the strands of the DNA with the number of full turns contained in the strands of a relaxed DNA molecule of the same length. Thus we introduce the following terminology.

**Definition.** Let $N$ denote the number of base pairs in a particular closed circular DNA molecule. Let $Lk_0 = N/10.5$ and define the *linking difference* to be $\Delta Lk = Lk - Lk_0$.

Because the linking number is a topological invariant, and $Lk_0$ will be a constant for a particular DNA molecule (independent of how supercoiled it is), the linking difference will also be a topological invariant. Note that, in general, $Lk_0$ will not be a linking number because it is not an integer. However, $Lk_0$ is roughly equal to the linking number of relaxed closed circular DNA of the same length.

Adding or subtracting a certain number of twists to a short relaxed DNA molecule will create more torsional stress than adding or subtracting the same number of twists to a long relaxed DNA molecule. Thus the maximum possible absolute value of the linking difference is larger for longer DNA molecules. As a way of normalizing the linking difference, we can also define the *specific linking difference* of a closed circular DNA molecule to be $\Delta Lk/Lk_0$.

Similarly, we define $Tw_0$ and $Wr_0$ to be the twist and average writhe, respectively, of a relaxed closed circular DNA molecule whose helix axis lies in a plane. Thus $Wr_0 = 0$ and $Tw_0 = Lk_0 = N/10.5$, where $N$ is the number of base pairs. We define the *writhe difference* $\Delta Wr = Wr - Wr_0 = Wr$ and the twist difference $\Delta Tw = Tw - Tw_0$. Then it follows from White's equation that $\Delta Lk = \Delta Tw + \Delta Wr$. Furthermore, Boles, White, and Cozzarelli (1990) have shown that the ratio of $\Delta Wr$ to $\Delta Tw$ is always approximately 2.6 : 1. This implies that $\Delta Wr = Wr$ is roughly $(0.72)\Delta Lk$ and $\Delta Tw$ is roughly $(0.28)\Delta Lk$. It follows that starting with relaxed DNA and introducing more or less linking will have a significantly greater impact on the average writhe than on the twist. Because the average writhe can be seen as the supercoiling of the DNA, this tells us that if we twist or untwist relaxed linear DNA before closing it up, it will generally cause the DNA to become supercoiled. For DNA in vivo, the linking difference will almost always be negative; in this case we say that the DNA is *negatively supercoiled*. In contrast, if the linking difference is positive we say that the DNA is *positively supercoiled*. When it is supercoiled, DNA is more compact than in its relaxed form. The larger the absolute value of the linking difference, the larger will be the absolute value of the writhing difference, the more supercoils there will be, and the more compact the molecule will be.

More compact DNA molecules can be experimentally distinguished from less compact DNA molecules by using a technique called *gel electrophoresis*. This technique involves placing a mixture of DNA molecules on one end of a strip of gel and applying different voltages to the two ends of the gel. This causes the more compact molecules to move across the gel faster than the less compact ones. In this way, molecular biologists can distinguish the linking numbers of different DNA molecules by determining how fast the different molecules move through gel electrophoresis.

## Toroidal Winding of DNA

So far, the supercoiling that we have been analyzing occurs when a DNA molecule winds around itself. This conformation of DNA is said to be *plectonemic* or *interwound*. Supercoiling can also occur when DNA is bound to a series of protein surfaces, causing the helix axis to lie on the surface of a torus. In particular, the helix axis will go once around the torus longitudinally, and possibly many times around meridionally. We can still use the concepts of Lk, Tw, and Wr, together with White's formula relating these concepts, in order to understand this type of toroidal winding. However, for DNA that is bound to protein surfaces, White, Cozzarelli, and Bauer (1988) have also developed an alternative way of splitting the linking number into two different quantities that are more closely related to the surface. These two quantities are the *surface linking number*, which characterizes how many times the helix axis wraps around the surface, and the *winding number*, which characterizes how many times the backbone of the DNA goes in and out of the surface.

In order to understand these concepts in more detail, we begin as before by orienting the DNA ribbon so that both strands and the helix axis have parallel orientations. Let $f$ denote the helix axis of the DNA and let $S$ denote an orientable surface that contains $f$. Choose a normal vector field to $S$. For some small $\varepsilon > 0$, we define a new oriented simple closed curve $f_\varepsilon$, by pushing $f$ a distance of $\varepsilon$ at each point in the direction of the normal vector field. The *surface linking number* of the DNA molecule, denoted by $S$Lk, is defined to be the linking number of the simple closed curves $f$ and $f_\varepsilon$.

As a simple example, if the surface $S$ is either a plane or a two-dimensional sphere, then $S$Lk $= 0$. If $S$ is a torus, we consider the solid torus $V$ that is bounded by $S$ in $\mathbb{R}^3$. Let $C$ denote the core of $V$, oriented so that it is parallel to the orientation of $f$. Because $f$ goes around once longitudinally, there is an ambient isotopy of $f$ in $V$ that takes the oriented curve $f$ to the oriented core $C$. The meridional twisting of $f$ around the torus $S$ does not prevent $f$ from being ambient isotopic to $C$. In contrast, because $f_\varepsilon$ was created by pushing $f$ in the direction of a normal vector field, all of the linking between $f$ and $f_\varepsilon$ is due to $f$'s wrapping meridionally around the torus $S$. So $SLk = Lk(f, f_\varepsilon) = Lk(f, C)$, and its absolute value represents the number of times that $f$ wraps around $S$ meridionally; furthermore, the sign of $SLk$ indicates the direction of the meridional wrapping. Because the linking number is always an integer, $SLk$ is always an integer and is relatively straightforward to compute. Also if the surface $S$ is smoothly deformed or if $f$ is deformed within the surface $S$, then $SLk$ will not change.

The winding number is closely related to the twist. In fact, if the helix axis lies in a plane, then the winding number of the DNA is identical to the twist. To define the winding number in general, we begin as we did previously for the twist. We parametrize the helix axis $f$ by a variable $t \in [0, 1]$; then for each value of $t$, we let $P(t)$ denote the plane perpendicular to $f$ at the point $f(t)$. Next we parametrize the backbone $g$ by $t$ in such a way that, for each $t$, the plane $P(t)$ intersects $g$ at $g(t)$. We define $v(t)$ to be the unit vector from $f(t)$ in the direction of $g(t)$. Let $u(t)$ denote the unit vector in the normal vector field to $S$ at the point $f(t)$. Then both $v(t)$ and $u(t)$ lie in the plane $P(t)$. We orient $P(t)$ so that $f(t)$ crosses $P(t)$ by going from the negative side of $P(t)$ to the positive side of $P(t)$. Let $\varphi(t)$ denote the angle measured from $u(t)$ to $v(t)$, as seen from the positive side of the plane $P(t)$. We want the winding number to represent the number of times that $v$ turns around $u$ as we go along $f$. That is, the winding number should be $1/2\pi$ times the total change in the angle $\varphi$. So we define the *winding number* to be $\Phi = 1/2\pi \int \varphi'(t)dt$.

Observe that because the DNA is closed, $v(0) = v(1)$ and $u(0) = u(1)$, so the total change in $\varphi$ must be a multiple of $2\pi$. It follows that $\Phi$ will always be an integer regardless of the surface $S$. Because we have defined $\varphi(t)$ as the angle from $u(t)$ to $v(t)$, as seen from the positive side of $P(t)$, we know that $\Phi$ will be positive if and only if the DNA backbone rotates in a right-handed helix around the helix axis. Also, note that if $S$ is smoothly deformed or the helix axis deforms itself within the surface, then the value of the winding number will not change.

The winding number also corresponds to half the number of times that the backbone passes through the surface, going from one side of the surface to the other side. This is because passing from one side of the surface to the other and then back to the first side corresponds to the vector $v(t)$ completing one full rotation about $u(t)$.

We would like to relate the various concepts that we have defined so far. In order to do this, it is convenient to define one more concept. The *surface twist*,

**Figure 7.8.** A ribbon with $SLk = 1$, $\Phi = 1$, and $Lk = 2$

$S$Tw, is defined to be the twist of $f_\varepsilon$ about $f$. White and Bauer (1988) proved that if Tw (as we defined it previously) is the twist of the backbone $g$ around the helix axis $f$, and $\Phi$ and $S$Tw are as defined above, then $Tw = S\text{Tw} + \Phi$. Combining this equation with White's earlier equation $Lk = \text{Tw} + \text{Wr}$ gives us the equation $Lk = S\text{Tw} + \Phi + \text{Wr}$. Now we can also apply the equation $Lk = \text{Tw} + \text{Wr}$ to the ribbon that has $f$ as its helix axis and $f_\varepsilon$ as a strand. However, the linking number of $f$ and $f_\varepsilon$ is by definition $SLk$. Also, the average writhe for the ribbon that has $f$ as its helix axis and $f_\varepsilon$ as a strand is equal to Wr for the DNA molecule, because the average writhe only depends on the helix axis and not on the backbone of the ribbon. Because $S$Tw is the twist of $f_\varepsilon$ about $f$, we have the equation $SLk = S\text{Tw} + \text{Wr}$. If we combine this with the equation $Lk = S\text{Tw} + \Phi + \text{Wr}$, which we obtained above, we get the equation $Lk = SLk + \Phi$. This equation is particularly nice because each of the quantities in it is an integer.

To get some intuition about this equation, we consider the simplest ribbon in which such an equation would be useful. Suppose we have a DNA molecule that has a helix axis $f$ lying on the surface of a torus and wrapping around once longitudinally, and once meridionally. Then consider a strand $g$ of the ribbon that winds once around the helix axis. So the strand runs parallel to the axis but goes once from the outside of the torus in $\mathbb{R}^3$ to the inside and then back out again (See Figure 7.8). As usual, we give the strand and the helix axis parallel orientations. We can see from the picture that $SLk = 1$ and $\Phi = 1$. Hence it follows from the above equation that the linking number of the helix axis and the strand will equal two.

We have seen that the linking number can be split into the surface linking number and the winding number, both of which are integers and not too hard to compute. It is also possible to determine all three of these quantities experimentally. Thus for DNA molecules whose helix axis lies on the surface of a torus, this equation can be a useful tool for understanding the topology and geometry of supercoiling.

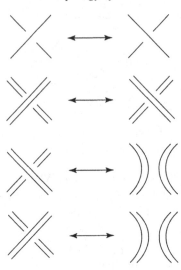

**Figure 7.9.** Some ways that enzymes can act on DNA

## The Action of Enzymes on DNA

When an enzyme acts on a DNA molecule, it can cause various changes in the structure of the molecule. One possible result is that one strand of duplex DNA will pass through the other strand. This has the effect of increasing or decreasing the value of Lk by one. This will, in turn, affect the value of Tw and the value of Wr so as to minimize the torsional stress on the molecule. Thus one result of the action of an enzyme is that the molecule will become more or less supercoiled. An enzyme can also act on pairs of strands, causing one or more pairs of strands to pass through each other, or one or more pairs of strands to be cut and reconnected another way. Some possible enzyme actions are illustrated in Figure 7.9.

Molecular biologists would like to understand the action of particular enzymes on the DNA in a cell. It is possible to experimentally determine the outcome of an enzyme action, but there is no known method to actually observe the action of an enzyme. Throughout the rest of this chapter we will see how knot theory can be used to model the actions of certain enzymes. Because DNA within cells is primarily linear, it may be difficult to detect the mechanism of an enzyme because any knotting or tangling can fall off the ends of linear DNA. In order to understand the mechanism of the enzyme, experiments are done with enzymes acting on closed circular DNA. The initial form of the DNA is called the *substrate*. Usually the substrate is unknotted, but this does not always have to be the case. When an enzyme acts on the substrate, it yields new forms of the DNA, which are called the *products* of the action. Gel electrophoresis is then used to separate molecules of different weights and shapes. In particular, it separates out knots and links according to how many crossings they have, because this corresponds to how compact the molecules are. The DNA is then coated with an *Escherichia coli* protein known as rec A. This makes the DNA fat enough for the crossings to be seen clearly with an electron microscope (see Krasnow et al., 1983 and Wasserman, Dungan, & Cozzarelli, 1985 for examples of such pictures). This technique of coating DNA with rec A was developed in 1983. Before that, it was not possible to see with an electron microscope which strands of the DNA contained overcrossings and which strands contained undercrossings. With the use of this technique it is now possible to determine exactly what kind of knots or links are contained in the product, and to use this to model the action of the enzyme.

Making use of the experimental findings of molecular biologists, we can translate the problem of understanding the actions of certain enzymes into a problem in knot theory. The general question is, if we know the knot or link types of the substrate and the products, can we figure out the action of the enzyme? Topologists Claus Ernst and DeWitt Sumners (1990) have answered this question for the enzyme TN3 Resolvase. We will present Ernst and Sumner's results after some preliminary background.

Formally, the term *recombination* is defined as the creation of new genetic sequences out of pieces of existing genetic sequences. It is an important part of reproduction because it enables the genetic material of two genetic parents to be brought together and combined in a different way to produce the genetic sequence of the offspring. Site-specific recombination involves two short linear pieces of duplex DNA with the same exact sequence of base pairs that are brought together by an enzyme, with the possible help of random thermal motion. The enzyme binds itself to the two short pieces that it has brought together (called *sites*), and then the enzyme cuts and recombines these two strands differently. This has the effect of changing the genetic code on the DNA.

Starting with supercoiled DNA, the result of recombination will be different depending on the exact nature of the supercoiling that is trapped between the two sites. Figure 7.10 illustrates three different possible outcomes for recombination, depending on the locations of two sites on a DNA molecule. Here we consider the duplex DNA molecule as a single ribbon, which we draw as a single simple closed curve, rather than as two separate strands that are twisted together. We have indicated the sites with arrows, which correspond to a particular orientation of the short sequence of base pairs along the site. During the recombination, the head of the first arrow is attached to the tail of the second

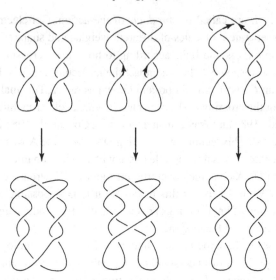

**Figure 7.10.** Different outcomes for recombination according to where the sites are

arrow, and the tail of the first arrow is attached to the head of the second arrow. The example in Figure 7.10 shows that the location of the sites makes a tremendous difference in the outcome of recombination.

An *electron micrograph* is the graphic image produced by an electron microscope. In such images, the part of the DNA that is bound by the enzyme is seen as a black disk with four strands coming out of it. Some enzymes bind to more than one pair of sites at a time, bringing about multiple recombination events simultaneously. Others bind only to two sites that are brought together in one spot. Sometimes the enzyme can cause more than one recombination event even at a single binding site.

There are two ways that two sites on a DNA molecule may line up. A particular orientation of the genetic code contained in each of the sites will induce an orientation on the entire circle. If the orientations induced on the entire circle by each of the oriented sites agree, then we say that the enzyme acts with *direct repeats*; otherwise we say that it acts with *inverted repeats*. In the case where the enzyme acts with direct repeats on a DNA molecule that has the form of a simple closed curve, and the enzyme acts at a single site an odd number of times, then the product will be an oriented link. In contrast, if the enzyme acts with direct repeats an even number of times, then the product will be an oriented knot. See, for example, Figure 7.11.

Molecular biologists Wasserman, Dungan, and Cozzarelli (1985) did experiments in which they started with unknotted, supercoiled, closed circular duplex

**Figure 7.11.** After one recombination event with direct repeats, we obtain a link; after two recombination events, we obtain a knot

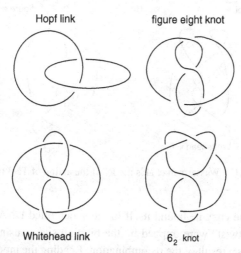

**Figure 7.12.** Outcomes after one, two, three, and four recombinations

DNA; they added the enzyme TN3 Resolvase and then analyzed the resulting product. The enzyme TN3 Resolvase is known to act on circular DNA with direct repeats, binding only at one spot. In approximately 95% of the cases, Resolvase brings about only one recombination event. However, in 5% of the encounters, it causes multiple recombination events at this same site. The quantity of products produced as a result of each of these recombinations decreases exponentially with the number of recombinations. So, when the quantity of each product is determined, the number of recombinations it took to produce each product can be determined. The results of the experiments with TN3 Resolvase acting on an unknotted circle yielded the results shown in Figure 7.12. The products were the Hopf link after one recombination event, the figure eight knot after two recombination events, the Whitehead link after three recombination events, and the $6_2$ knot after four recombination events. After five rounds of recombination, if anything was produced, it was too minute to detect.

Wasserman et al. used these experimental results to make a model of how they believed the enzyme was acting. Their model, illustrated in Figure 7.13,

214                         7. *Topology of DNA*

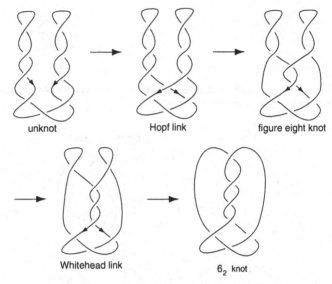

unknot              Hopf link          figure eight knot

Whitehead link              6₂ knot

**Figure 7.13.** Wasserman et al.'s model of the action of TN3 Resolvase

suggests that the enzyme bound itself to the supercoiled DNA in such a way
that three half-twists were trapped by the binding site, causing the links and
knots to form as a result of the recombination. Reading the model from the left
to the right, we can see the products after each successive recombination event.

## Conway's Theory of Tangles

Using the theory of tangles, Ernst and Sumners (1990) were able to prove that
Wasserman et al.'s model was correct. In order to explain Ernst and Sumner's
work we must first learn about tangles. The theory of tangles was originally
developed by Conway (Conway, 1967), and Ernst and Sumner's results build
on Conway's work. We begin with some definitions.

By a *three-dimensional ball* we shall mean the unit ball $\{(x, y, z) \in \mathbb{R}^3 \mid x^2 + y^2 + z^2 \leqslant 1\}$. This three-dimensional ball intersects the $xy$ plane in a disk. We
can imagine the positive $y$ axis as pointing north and the positive $x$ axis as
pointing east. Thus we shall label points on the boundary of this disk by the
pairs of letters $\{NE, NW, SW, SE\}$ as we would label points on a compass.

**Definition.** A *tangle* $(B, t)$ is a three-dimensional ball $B$, containing two disjoint arcs together with a finite number (possibly zero) of disjoint simple closed
curves that are all represented by $t$, such that the intersection of $t$ with the

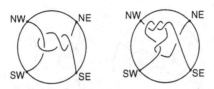

**Figure 7.14.** Some tangles

boundary of $B$ is precisely the set of endpoints of the two arcs. Furthermore, there is a fixed orientation-preserving homeomorphism from $B$ to the unit three-dimensional ball that takes the endpoints of the arcs to the specified points {NE, NW, SW, SE}.

In Figure 7.14 we illustrate two different tangles projected onto the disk where the $xy$ plane meets the three-dimensional ball. Normally, we do not label the compass points because it is understood that the upper left is NW, and so on.

Just as we wanted to know what it means for two knots to be equivalent, we shall want to know what it means for two tangles to be equivalent.

**Definition.** Two tangles $(B, t_1)$ and $(B, t_2)$ are said to be equivalent if there is an orientation-preserving homeomorphism $h : (B, t_1) \to (B, t_2)$ that is the identity on the boundary of $B$.

Building on Reidemeister's Theorem from Chapter 2 (Reidemeister, 1926), we could show that two tangles are equivalent if and only if we can go from a projection of one to a projection of the other by using a finite sequence of Reidemeister moves that do not move the endpoints of the arcs.

There are two types of operations that we can do on tangles that we describe below. We can add two tangles $A$ and $B$ to obtain a new tangle $A + B$ by first shrinking $A$ and $B$ so that they are each approximately one-fifth of their original size, and then placing the shrunken $A$ and $B$ inside of a unit three-dimensional ball and connecting the balls with arcs as illustrated in Figure 7.15. Thus $A + B$ is a tangle according to our definition. We can also form a knot or a link from a tangle $A$ by add an arc on the top and the bottom of the tangle as illustrated on the right side of Figure 7.15. The knot or linked formed in this way is denoted by $N(A)$ and is said to be the closure of $A$. We can combine these two operations by adding tangles $A$ and $B$ and then taking the closure to obtain $N(A + B)$. In Exercise 19 you will prove that tangle addition is associative; that is, $N[(A + B) + C] = N[A + (B + C)]$. Because of this equality we can write $N(A + B + C)$ without any ambiguity.

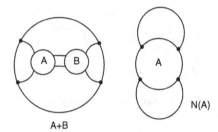

**Figure 7.15.** The sum and closure operations

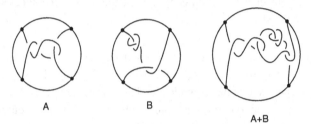

**Figure 7.16.** The sum of two tangles

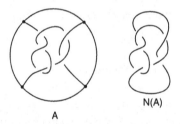

**Figure 7.17.** The closure of a tangle

Figure 7.16 provides an example of how we can take the sum of two tangles to obtain a new tangle, and Figure 7.17 illustrates the closure of a tangle. Because taking the closure of a tangle always yields a knot or a link, we see that the theory of tangles is closely related to the theory of knots and links.

We are interested in a special class of tangles called *rational tangles*. Rational tangles are useful because they are well understood, and in addition they are made up of horizontal and vertical twists, so they resemble what molecular biologists see when they look at electron micrographs of DNA molecules.

**Definition.** A tangle $(B, t)$ is said to be *rational* if $t$ can be deformed by an ambient isotopy in $B$ to the trivial tangle (illustrated in Figure 7.18), where the endpoints of $t$ remain in the boundary of $B$ throughout the isotopy.

**Figure 7.18.** The trivial tangle

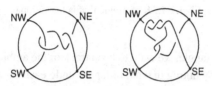

**Figure 7.19.** Some rational tangles

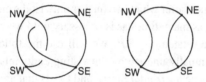

**Figure 7.20.** As we progressively untwist the tangle on the left in Figure 7.19, we get these two tangles

Note that the ambient isotopy in this definition does not necessarily fix every point in the boundary. Intuitively, a tangle is rational if it can be untwisted by moving the endpoints along the boundary of the ball $B$. Consider, for example, the tangles illustrated in Figure 7.14, which we redraw in Figure 7.19. These tangles are both rational.

We can see as follows how to untwist the tangle on the left in Figure 7.19 to obtain the trivial tangle. In order to untwist the right side of the tangle, we move the two endpoints SE and NE around in a full clockwise circle in the Eastern hemisphere to obtain the link that is on the left in Figure 7.20. Then we move the two endpoints, which are now in positions SW and SE, around in a full counterclockwise circle in the Southern hemisphere to obtain the link on the right in Figure 7.20. Finally, we can obtain the trivial tangle by rotating this latter tangle by 180°.

In contrast with the tangles in Figure 7.19, Figure 7.21 provides an example of a tangle that is not rational. Observe that no matter how we move the arcs in the tangle, if we keep the endpoints of the arcs in the boundary of $B$, then we will not be able to remove the knot in one of the arcs or separate the two arcs.

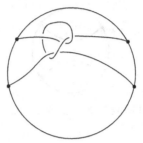

**Figure 7.21.** A tangle that is not rational

For any rational tangle, we want to keep track of the ambient isotopy that we use to untangle it by noting in what order we undo each of the different sets of twists. We shall always undo our rational tangles by starting with the right side of the tangle and untwisting the endpoints labeled SE and NE some number of times (possibly zero). Then we shall untwist the endpoints labeled SE and SW some number of times, after which we further untwist the endpoints labeled SE and NE, and so on, as many times as is necessary until we get the trivial tangle or the trivial tangle rotated by 180°. We will use the following terminology. Twists that are eventually undone by untwisting endpoints in the SE and NE positions are said to be *horizontal twists*, and twists that are eventually undone by untwisting the SE and SW endpoints are said to be *vertical twists*. A twist is said to be *right handed* if you could slide your right thumb along one string and your right forefinger along the other string as the strings twist around one another horizontally or vertically, and a twist is said to be *left handed* if you could slide your left thumb along one string and your left forefinger along the other string as the strings twist around one another. A right-handed twist is considered to be *positive*, and a left-handed twist is considered to be *negative*. For example, when we untwisted the tangle on the left side of Figure 7.19, we first undid two positive horizontal half-twists and then we undid two negative vertical half-twists. In order to untwist the tangle on the right side of Figure 7.19, we first undo one positive horizontal half-twist, and then we undo two positive vertical half-twists, and finally we undo three negative horizontal half-twists.

In general, every rational tangle can be specified by a vector $(a_1, a_2, \ldots, a_n)$ with integer entries that tells us how to untwist the tangle to get to the trivial tangle or the trivial tangle rotated by 180°. Starting with a rational tangle, we first undo $a_n$ horizontal half-twists by moving the endpoints SE and NE (where $a_n$ may be zero), then we undo $a_{n-1}$ vertical half-twists by moving the endpoints SW and SE, and so on until we undo $a_1$ either vertical or horizontal half-twists, depending on whether $n$ is even or odd, respectively. After undoing all of these half-twists we end up with either the trivial tangle, which is denoted by the vector (0), or the trivial tangle rotated by 180°, which is denoted by the vector (0, 0). Note that we require that $a_n$ must denote some number of horizontal twists, even if this number is zero, while all other entries must be nonzero. This is so that we will know whether we begin by undoing horizontal twists or vertical twists. For example, the first tangle in Figure 7.19 has vector $(-2, 2)$ and the second tangle in Figure 7.19 has vector $(-3, 2, 1)$.

The reason that rational tangles are called rational tangles is that from every vector representing a rational tangle we can associate a unique rational number $\beta/\alpha$ (here we are also allowing $\infty = 1/0$ as a rational number), by using continued fractions as follows. We define $\beta/\alpha$ by the property that

$$-\beta/\alpha = a_n - \cfrac{1}{a_{n-1} - \cfrac{1}{a_{n-2} - \cfrac{1}{\cdots \cfrac{1}{a_2 - \cfrac{1}{a_1}}}}}$$

Building on the work of Schubert (1956), Conway (1967) proved that two rational tangles are equivalent if and only if their associated rational numbers are the same.

## Results of Ernst and Sumners

We will now see how the theory of tangles enabled Ernst and Sumners (1990) to analyze the action of the enzyme TN3 Resolvase on DNA molecules. They considered the unknotted substrate DNA molecule as the closure of the sum of two tangles. One tangle, called the *site tangle* and designated by $T$, is the tangle that is bound inside of the enzyme ball before the enzyme acts. The other tangle consists of whatever tangling is outside of the enzyme. This second tangle is called the *substrate tangle* and is designated by $S$. A single recombination event changes the tangle $T$ into a new tangle, which is now inside the enzyme ball and is now called the *recombination tangle*; it is designated by $R$. Thus the original unknotted substrate has the form $N(S + T)$ and the product after a single recombination event has the form $N(S + R)$ because the enzyme has replaced the tangle $T$ by the tangle $R$. Molecular biologists make the assumption that a given enzyme always acts the same way. In particular they assume that, if the enzyme causes multiple recombination events within the same ball, then each of these events is the same. For Ernst and Sumner's model this assumption is interpreted to mean that the product after two recombination events is

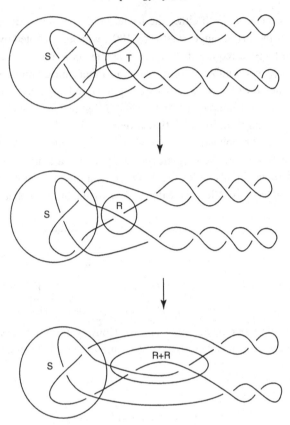

**Figure 7.22.** The tangles $S$, $T$, and $R$ in the model of the action of TN3 Resolvase

$N(S + R + R)$ and an additional $R$ is added to the tangle after each subsequent recombination event. In Figure 7.22, we use the model of Wasserman et al. (1985) to illustrate the tangles $S$, $T$, and $R$. Observe that we do not include the supercoiling to the right of the enzyme ball in the tangle $S$. These twists can be removed by a deformation of the knot or link while the enzyme ball is kept fixed. Hence, these twists have no topological significance.

Ernst and Sumners (1990) proved the following theorem. See Figure 7.12 for illustrations of the links involved. For those readers who wish to read the work of Sumners and Ernst, we note that our sign conventions for the vector representing a tangle are different than theirs.

**Theorem 7.1.** *Suppose that* S, T, *and* R *are tangles satisfying the following equations:* (1) N(S + T) = *the unknot;* (2) N(S + R) = *the Hopf link;*

(3) N(S + R + R) = *the figure eight knot; and* (4) N(S + R + R + R) = *the Whitehead link. Then* S *and* R *are both rational tangles,* S = (3, 0), R = (1), *and* N(S + R + R + R + R) *is the* $6_2$ *knot.*

Theorem 7.1 mathematically proves what Wasserman et al. had suggested in their original model of the action of TN3 Resolvase. That is, as illustrated in Figure 7.22, *S* must contain exactly three vertical right-handed twists, and *R* must contain one horizontal right-handed twist. Theorem 7.1 also correctly predicts that the outcome of four recombinations will be the $6_2$ knot. Note that the theorem cannot tell us anything about the amount of topologically trivial supercoiling, which is not contained in any of the tangles.

In order to prove Theorem 7.1, we need to understand the relationship between rational tangles and a particular type of knots and links that we define below.

**Definition.** A knot or link $L$ is said to be a *two-bridge* knot or link if there exists a rational tangle $A$ such that $L = N(A)$. A knot or link $L$ is said to be *Montesinos* if there exist rational tangles $A_1, A_2, \ldots, A_n$ such that $L = N(A_1 + A_2 + \cdots + A_n)$.

Observe that this definition does not say that if $L$ is a two-bridge link then for any tangle $A$ such that $L = N(A)$ then $A$ must be rational, or the analogous statement for Montesinos links. The set of two-bridge links is a subset of the set of Montesinos links. With a little bit of effort the reader should be able to see that all of the products obtained by TN3 Resolvase are actually two-bridge links (see Exercise 17). For biological reasons we expect tangles $S$ and $R$ to be made up of horizontal or vertical twists, and hence we expect them to be rational. Thus we expect all the products to be Montesinos knots and links.

In Chapter 4 we defined the connected sum of two knots. Now we want to extend this notion to tangles as well. In particular, recall the following definition from Chapter 4.

**Definition.** Let $K$ be a knot in $S^3$ or $\mathbb{R}^3$. Suppose that there exists a surface $F$ homeomorphic to a two-dimensional sphere such that $F$ meets $K$ in two points $p$ and $q$. Let $B$ be an arc in $F$ with endpoints $p$ and $q$. Consider the simple closed curves obtained by joining each of the components of $K - \{p, q\}$ to the arc $B$. If neither of these simple closed curves is the unknot, then we say that $K$ is a *connected sum*.

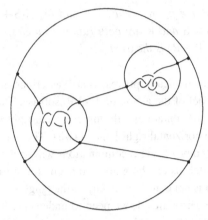

**Figure 7.23.** The connected sum of a tangle and a link

We can take the connected sum of a tangle with a knot or a link by adding to the tangle a ball that contains a knotted arc or an arc that is linked to a simple closed curve. For example, see Figure 7.23.

Formally, we have the following definition.

**Definition.** Let $(B, t)$ be a tangle and suppose that there exists a surface $F$ homeomorphic to a two-dimensional sphere such that $F$ is contained in $B$ and $F$ meets $t$ in two points $p$ and $q$. Let $A$ be an arc in $F$ with endpoints $p$ and $q$. Let $B'$ denote the ball in $B$ that is bounded by $F$. Consider the knot or link $L$ obtained by joining the part of $t$ that is in $B'$ to the arc $A$. Let $(B, s)$ denote the tangle obtained from $(B, t)$ by replacing the part of $t$ that is in $B'$ by the arc $A$. Then we say that $(B, t)$ is the *connected sum* of $(B, s)$ and $L$, and we say that $L$ is a *connected summand* of $(B, t)$.

Observe that in the definition of the connected sum of two knots, we require that each of the two knots be nontrivial. In contrast, for tangles we can speak of the connected sum of a tangle $(B, t)$ and a trivial knot. Of course, such a connected sum just yields our original tangle $(B, t)$. Also note that the connected sum of a tangle and a knot or a link should not be confused with the sum of two tangles $A + B$ that we introduced in Figure 7.16.

We shall use the following lemma about tangles and two-bridge links, which follows from the work of Bill Menasco (1984) on alternating links. We do not provide a proof of these results. The first part of the lemma follows from Menasco's Theorem 3b and its proof, and the second part can be obtained by using an argument that is similar to the proof of Menasco's Theorem 3b.

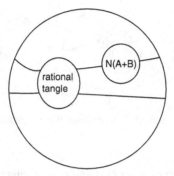

**Figure 7.24.** The connected sum of a rational tangle and the knot $N(A + B)$

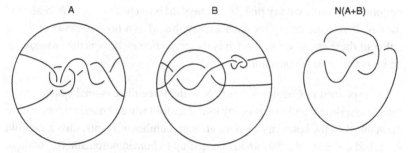

**Figure 7.25.** An example in which neither $A$ nor $B$ is rational, and $B$ is the connected sum of a rational tangle with $N(A + B)$

**Lemma 7.2.** *Let* A *and* B *be tangles.* (1) *If* $N(A + B)$ *is a two-bridge link, then at least one of* A *and* B *is either a rational tangle or the connected sum of a rational tangle with* $N(A + B)$ *(as in Figure* 7.24). (2) *If* $A + B$ *is a rational tangle, then either both* A *and* B *are rational tangles or* $N(A + B)$ *is the unknot.*

We give examples of the two parts of Lemma 7.2 below. Consider tangles $A$ and $B$ that are illustrated in Figure 7.25. We can see that $N(A + B)$ is a trefoil knot (the reader should check this). The tangle $B$ is the connected sum of a rational tangle with $N(A + B)$, and it can be shown that neither $A$ nor $B$ is a rational tangle.

Figure 7.26 gives an example of part 2 of Lemma 7.2.

A key ingredient in the proof of Theorem 7.1 is the use of twofold branched covers. We shall use the following definition from Chapter 3.

**Definition.** Let $M$ and $N$ be three-manifolds, and let $h : M \to M$ be an orientation-preserving homeomorphism with an order of two. Let $p : M \to N$

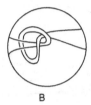 

A                           B                      A+B

**Figure 7.26.** A rational tangle $A + B$ in which $B$ is not rational, and $N(A + B)$ is the unknot

be defined such that $p(x) = p(y)$ if and only if either $x = y$ or $h(x) = y$. Suppose that $p$ is a continuous onto map that takes open sets to open sets. Let $A$ denote the set of points $x$ in $M$ such that $h(x) = x$. If $B = p(A)$ is a one-manifold, then we say that $M$ is a *twofold branched cover* of $N$ branched over $B$. If $A$ is the empty set, then we say that $M$ is a *twofold cover* of $N$. In either of these cases, we say that $p$ is the *projection map*, $N$ is the *base space*, and $h$ is the *covering involution*.

We explained in Chapter 3 that if $N$ is the three-dimensional sphere or the three-dimensional ball and $B$ is any one-manifold whose boundary meets each component of the boundary of $N$ in an even number of points, then a twofold branched cover of $N$ exists and is unique up to homeomorphism. So, without ambiguity, we can speak of the twofold branched cover of $N$ branched over $B$. In Chapter 3, we saw that the twofold branched cover of $S^3$ branched over the unknot is $S^3$, and we saw how to construct the twofold branched cover of $S^3$ branched over a nontrivial knot or link. Now, starting with a tangle $(B, t)$, we would like to understand the twofold branched cover of the ball $B$ branched over $t$. If $A$ represents a tangle $(B, t)$, then we refer to the twofold branched cover of $B$ branched over $t$ as the *twofold branched cover of A*.

We shall first construct the twofold branched cover of a trivial tangle $(B, t)$. Let $D$ be a disk that divides $B$ in half such that one arc of $t$ is in each component of $B - D$ (see Figure 7.27).

Let $E$ and $F$ be the components of $B - D$, let $p$ be the arc of $t$ that is contained in $E$, and let $q$ be the arc of $t$ that is contained in $F$. Then $E$ and $F$ are each three-dimensional balls. We shall construct the twofold branched cover of the ball $E$ branched over the arc $p$. Note that $E$ can actually be deformed to a round ball where $D$ is a curved disk in its boundary. Let $S$ denote a surface that is homeomorphic to a disk and that is bounded by $p$ together with an arc of $E$ in the boundary of $B$. We cut $E$ open along $S$ and let $Y$ denote the cut open ball with a copy of $S$ on each side of the cut where $S$ was. We label one copy as $S_+$ and the other as $S_-$. Thus $Y$ is a ball whose boundary can be thought of

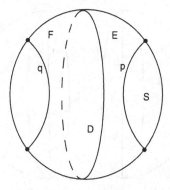

**Figure 7.27.** A trivial tangle separated by a disk

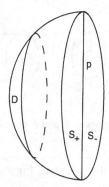

**Figure 7.28.** We split $E$ open along $S$ to get $Y$

as a hemisphere together with the two copies of $S$, which are attached together along $p$. The disk $D$ is contained in the hemispheric part of the boundary of $Y$. In Figure 7.28 we have flattened the two copies of $S$ and rounded the rest of the boundary of $Y$.

Now let $Y_1$ and $Y_2$ denote copies of $Y$, and let $U$ be the ball obtained by gluing together $Y_1$ and $Y_2$, where $S_+$ of $Y_1$ is glued to $S_-$ of $Y_2$ and $S_-$ of $Y_1$ is glued to $S_+$ of $Y_2$. The ball $U$ is the twofold branched cover of $E$ branched over $p$, where the covering involution rotates $U$ by $180°$ around $p$, interchanging $Y_1$ and $Y_2$ and interchanging the two copies of $S$ (see Figure 7.29). By an identical construction we obtain a ball $V$, which is the twofold branched cover of $F$ branched over the arc $q$.

Now each of the balls $U$ and $V$ contains two copies of the disk $D$, which we shall now call $D_1$ and $D_2$. We can deform $U$ and $V$ to two solid cylinders with the arcs $p$ and $q$, respectively, running through their centers, and $D_1$ and $D_2$

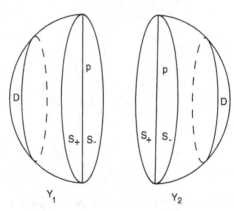

**Figure 7.29.** We glue $Y_1$ and $Y_2$ together to get the twofold cover of $E$ branched over $p$

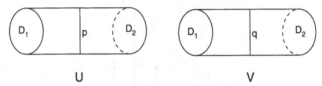

**Figure 7.30.** The twofold coverings of $E$ and $F$ branched along $p$ and $q$, respectively

are the disks on the ends of the cylinders. Figure 7.30 illustrates $U$ and $V$ in this form.

Because $E$ and $F$ were originally glued together along $D$, in order to obtain the twofold branched cover of $B$ branched over $t$, we glue together $U$ and $V$ along the pairs of disks $D_1$ and $D_2$. Thus the twofold branched cover of $B$ branched along $t$ is a solid torus, where the covering involution is a rotation by $180°$ around an axis that cuts through the solid torus in the two arcs $p$ and $q$. This rotation takes $U$ to itself and takes $V$ to itself, and it exchanges the disks $D_1$ and $D_2$. Figure 7.31 illustrates this solid torus together with the covering involution.

Because a rational tangle is defined to be a tangle that is ambient isotopic to a trivial tangle, every rational tangle is homeomorphic to a trivial tangle. Hence the twofold branched cover of a rational tangle is also a solid torus. Observe that, up to conjugation by a homeomorphism, the order-two homeomorphism illustrated in Figure 7.31 is the only order-two homeomorphism of a solid torus that fixes two arcs. Thus a rational tangle is the only tangle whose twofold branched cover is a solid torus. It follows that a tangle is rational if and only if its twofold branched cover is a solid torus.

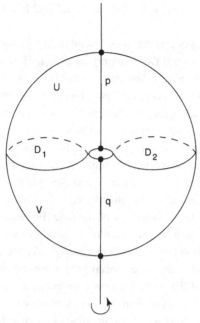

**Figure 7.31.** The twofold covering of $B$ branched along $t$ is a solid torus

Now suppose that we wish to construct the twofold branched cover of a tangle $(A, s)$ that is not rational. Observe that just as we would obtain $S^2$ by gluing two disks together along their boundaries, we will obtain $S^3$ by gluing two balls together along their boundaries. Furthermore, by gluing the tangle $(A, s)$ to the trivial tangle $(B, t)$ (or the trivial tangle rotated by 180°) along the boundaries of the balls $A$ and $B$, we obtain some (possibly trivial) knot or link $K$ in $S^3$. In Chapter 3, we saw how to find the twofold branched cover $M$ of $S^3$ branched over $K$. We let $M$ be the three-manifold obtained by gluing together the twofold branched cover of $A$ branched over $s$ and the twofold branched cover of $B$ branched over $t$, along their boundaries. Then $M$ is the twofold branched cover of $S^3$ branched over a knot or link. Because $(B, t)$ is trivial, we know that the twofold branched cover of $(B, t)$ is a solid torus. Thus we can obtain the twofold branched cover of $(A, s)$ by removing this solid torus from $M$. Observe that whether or not a tangle $(B, t)$ is rational, the boundary of the twofold cover of $B$ branched over $t$ will be a torus.

With this background in hand, it is possible to understand the ideas of the proof of Theorem 7.1. However, the proof itself is quite technical, so we shall give a sketch of the proof rather than presenting the proof in complete detail. Even this description of the proof is difficult and requires some experience with low dimensional topology, so the reader should feel free to skim it or to skip over it.

The proof uses classical techniques of knot theory, including the concepts of the fundamental group and Dehn surgery on a knot. The fundamental group is an important topological tool; however, it would take us too far afield to define it here. For the reader who is unfamiliar with the fundamental group, it can be thought of as a particular group that we associate with a three-manifold such that if two three-manifolds are homeomorphic then they have the same fundamental group. Thus it is a topological invariant. Although most fundamental groups are infinite, there are some that are finite. Dehn surgery on a knot involves removing a knotted solid torus from $S^3$ and then gluing this solid torus back along its boundary in such a way that the boundary of the solid torus may be attached differently to the boundary of the hole. For example, we could glue back the solid torus so that a meridian of the solid torus is now glued where a curve going once around the torus longitudinally and once around meridionally was previously. In addition to the concepts of the fundamental group and Dehn surgery, the proof of Theorem 7.1 uses a deep result of Culler et al. (1987) called the Cyclic Surgery Theorem. For our purposes we only need a simplified version of the Cyclic Surgery Theorem, which says that if two different Dehn surgeries on a knot both yield three-manifolds that have finite fundamental groups and if the orders of these two groups differ by more than one, then the knot on which we did Dehn surgery is either a trivial knot or a torus knot.

*Sketch of the Proof of Theorem 7.1.* The main part of the proof is to show that both of the tangles $S$ and $R$ must be rational. To do this we need to apply Lemma 7.2 several times as follows.

To begin, we suppose that $R$ is not a rational tangle. By part 2 of Lemma 7.2, if $S + R$ were rational then $N(S + R)$ would be the unknot because $R$ is not rational. However, $N(S + R)$ is the Hopf link, so $S + R$ cannot be rational. Now $N[(S + R) + R]$ is the figure eight knot, which can easily be shown to be a two-bridge link (the reader should show this). Because neither $S + R$ nor $R$ is rational, part 1 of Lemma 7.2 shows that either $S + R$ or $R$ must contain the figure eight knot as a connected summand. In either case, this would imply that $N(S + R)$ contains the figure eight knot as a connected summand. Because $N(S + R)$ is the Hopf link, both components of which are unknotted, this is impossible. This contradiction proves that $R$ must be rational.

Now we suppose that $S$ is not rational. Suppose also, for the moment, that $R + R$ is not rational. By associativity of tangle addition, $N[S + (R + R)] = N[(S + R) + R]$ is the figure eight knot. So by part 1 of Lemma 7.2, either $S$ or $R + R$ contains the figure eight knot as a connected summand. This implies that

$N(S + R + R + R)$ also contains the figure eight knot as a connected summand. This is impossible because $N(S + R + R + R)$ is the Whitehead link, both components of which are unknotted. Thus if $S$ is not a rational tangle, then $R + R$ must be a rational tangle.

Similarly, because $N(S + T)$ is the unknot (which is a two-bridge knot), by part 1 of Lemma 7.2, either $S$ or $T$ is rational or is a connected sum of a rational tangle with $N(S + T)$. Because $S$ is not rational by assumption, and $N(S + T)$ is the unknot, $T$ is either rational or is the connected sum of a rational tangle with the unknot. Because taking the connected sum with the unknot has no effect, in either case, $T$ is rational. So if $S$ is not rational then both $R + R$ and $T$ must be rational.

Next we consider the twofold branched covers of $S^3$ branched over $N(S + T)$, $N(S + R)$, and $N(S + R + R)$, which we denote by $X$, $Y$, and $Z$, respectively. The twofold branched cover of the unknot is $S^3$, so we know that $X$ is $S^3$. We denote the twofold branched covers of the tangles $T$, $R$, $R + R$, and $S$ by $T'$, $R'$, $(R + R)'$, and $S'$, respectively. Because $T$, $R$, and $R + R$ are rational, the twofold branched covers $T'$, $R'$, and $(R + R)'$ must all be solid tori.

Recall that $S^3$ is the union of two balls glued along their boundaries. Also, in Exercise 20 you will prove that we can obtain the knot or link $N(S + T)$ by gluing the tangles $S$ and $T$ together along the boundaries of their balls. Thus $X$, the twofold branched cover of $S^3$ branched along $N(S + T)$, is homeomorphic to the union of $S'$ and $T'$ glued along their torus boundaries. We know that $X$ is the three-dimensional sphere and $T'$ is a solid torus. However, the complement of any solid torus in $S^3$ is a (possibly trivial) knot complement. So $S'$ is a knot complement. Similarly, we can think of $Y$ as the union of $S'$ and $R'$ glued along their torus boundaries, and $Z$ as the union of $S'$ and $(R + R)'$ glued along their torus boundaries. Because $X$, $Y$, and $Z$ all contain $S'$, we see that each of $Y$ and $Z$ can be obtained from $X$ by replacing the solid torus $T'$ by the solid tori $R'$ and $(R + R)'$, respectively, attached appropriately. Because all solid tori are homeomorphic, $Y$ and $Z$ are each obtained from $X = S^3$ by removing the solid torus $T'$ and sewing it back differently. By definition this is Dehn surgery on the knot whose complement is $S'$. The twofold branched cover of the Hopf link is known to have fundamental group $\mathbb{Z}_2$, which has an order of two, and the twofold branched cover of the figure eight knot is known to have fundamental group $\mathbb{Z}_5$, which has an order of five (see Burde & Zieschang, 1985). The orders of these groups differ by more than one, so the Cyclic Surgery Theorem (Culler et al., 1987) implies that $S'$ must be either the complement of a torus knot or the complement of a trivial knot. Results about torus knots (Moser, 1971) now imply that $S'$ is in fact a solid torus. It follows that $S$ is rational, contradicting our assumption. This proves that $S$ as well as $R$ must be a rational tangle.

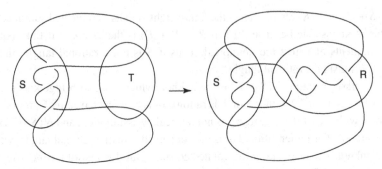

**Figure 7.32.** If $R = (3)$ then we can obtain the Whitehead link after one recombination event

It follows that $N(S + R)$, $N(S + R + R)$, and $N(S + R + R + R)$ are all written in the form of Montesinos links. In the early 1970s, Montesinos (1973) classified all such links. By our hypotheses $N(S + R)$ is the Hopf link, $N(S + R + R)$ is the figure eight knot, and $N(S + R + R + R)$ is the Whitehead link. The classification of Montesinos links (see Burde & Zieschang, 1985) together with a computation implies that $S = (3, 0)$ and $R = (1)$. It follows easily that $N(S + R + R + R + R)$ is the $6_2$ knot.  □

After having developed their model of TN3 Resolvase, Wasserman et al. (1985) wondered whether it was possible for an enzyme to change an unknot into a Whitehead link by means of one recombination event. Using the terminology of Theorem 7.1, we see that we could obtain the Whitehead link after one recombination event, for example, if $S = (3, 0)$, $T = (0)$, and $R = (3)$. We illustrate this in Figure 7.32.

However, molecular biologists believe that every recombination event must actually have $T = (0)$ and $R = (\pm 1)$ (Ernst & Sumners, 1990). Under these biological assumptions, Ernst and Sumners answered the above question in the negative. Specifically, they proved the following theorem (Ernst & Sumners, 1990) by using methods similar to those in the proof of Theorem 7.1.

**Theorem 7.3.** *There is no tangle* S *such that both* N[S + (0)] *is the unknot, and* N[S + (±1)] *is the Whitehead link.*

Ernst and Sumners have continued their work using tangles to understand the action of enzymes on DNA. They would like to be able to use tangles to model the action of various enzymes as they did for TN3 Resolvase. With this goal in mind, Ernst and Sumners (1999) recently proved the following general theorem. Note that a tangle is said to be an *integral tangle* if consists entirely of horizontal twists, with no vertical twists.

**Theorem 7.4.** *Let $L_0$ be a two-bridge knot or link, and let $L_1$, $L_2$, and $L_3$ be two-bridge knots or links that are not all the same. Suppose that for $0 \leqslant n \leqslant 3$, we have the equations $N(S + nR) = L_n$. Then there is at most one pair of tangles $\{S, R\}$ that satisfies all three of these equations. Furthermore, if such a pair of tangles exists, then S must be either a rational tangle or the sum of two rational tangles, R must be an integral tangle, and at least one of $L_0$, $L_1$, $L_2$, and $L_3$ must be topologically chiral.*

In the hypothesis of Theorem 7.1, we had assumed that $N(S + T)$ is the unknot, $N(S + R)$ is the Hopf link, $N(S + R + R)$ is the figure eight knot, and $N(S+R+R+R)$ is the Whitehead link. If we assume that $T$ is the trivial tangle (which makes sense biologically), then $N(S + T) = N(S) = N[S + 0(R)]$ will also be the unknot. We can let $L_0$ be the unknot, $L_1$ be the Hopf link, $L_2$ be the figure eight knot, and $L_3$ be the Whitehead link. Then the equations in the hypotheses of Theorem 7.1 satisfy the hypotheses of Theorem 7.4. Furthermore, the conclusion of Theorem 7.1 that $S = (3, 0)$ and $R = (1)$, together with the observation that the Whitehead link is topologically chiral, is consistent with the more general conclusion of Theorem 7.4. It follows that Theorem 7.4 is, in fact, a generalization of Theorem 7.1.

## Exercises

1. Give an example not included in the text of each of the following:
   a) How the location of the pair of sites on unknotted closed circular DNA that are brought together by the enzyme can affect the products of site-specific recombination.
   b) A rational tangle and its vector.
   c) A non-rational tangle.
2. Explain how Theorem 7.1 says that Wasserman was correct in his model of the action of TN3 Resolvase.
3. Prove that closed circular duplex DNA always contains an even number of half-twists.
4. For the ribbon on the right in Figure 7.7, is the writhe always equal to $-1$ no matter how we project the helix axis, or is there some well-defined projection where the writhe is not equal to $-1$? Explain.
5. Use the formal definition of the twist to explain why Tw $\approx 0$ for the ribbon on the right of Figure 7.7.
6. Assume that the two ribbons in Figure 7.7 each contain 105 base pairs. Decide whether each of these two ribbons is negatively supercoiled, positively supercoiled, or neither. Explain.

7. Explain why the winding number is positive if and only if the DNA backbone rotates in a right-handed helix around the helix axis.

8. Suppose that the helix axis of a closed circular duplex DNA molecule lies on the surface of a torus and wraps once around longitudinally and twice around meridionally. (a) Suppose that a strand of the DNA winds three times around the helix axis. Find $SLk$, $\Phi$, and Lk. (b) Suppose that a strand of the DNA winds three times around the helix axis in the opposite direction as in part a. Find $SLk$, $\Phi$, and Lk.

9. Draw a picture of a ribbon that has Lk $= 0$, Wr $\approx 2$, and Tw $\approx -2$.

10. In each of the examples in Figure 7.10, is the enzyme acting with direct repeats or inverted repeats?

11. Explain why if the enzyme acts an even number of times with direct repeats then the product must be a single simple closed curve. What happens when an enzyme acts an odd number of times with inverted repeats? Explain.

12. Draw the following tangles: $(1)$, $(-1)$, $(1, 0)$, $(-1, 0)$, $(3, 2, 1)$, $(3, 2, 1, 0)$, and $(1, 2, 3, 0)$.

13. Let $\beta = 2$ and $\alpha = 3$; then write the tangle vector associated with $\beta/\alpha$ and draw the tangle that has that vector.

14. For every tangle $A$ and $B$, is $N(A+B)$ necessarily equivalent to $N(B+A)$? Explain.

15. If $A$ and $B$ are both rational tangles, is $A + B$ necessarily a rational tangle? Explain.

16. Suppose that $A$ is a rational tangle, and $A = (a_1, \ldots, a_n)$ where the $a_i$ have alternating signs. Prove that $N(A)$ is an alternating link. (See Chapter 2 for the definition of an alternating link.)

17. Prove that for every rational tangle $A$ there is a rational tangle $B$ such that $N(A+B)$ is the trivial knot. Is this true if the tangle $A$ is not rational? Explain.

18. Draw a tangle $A$ that is not rational, such that $N(A)$ is a two-bridge link with two components. Draw a tangle $B$ that is not rational, such that $N(B)$ is a nontrivial two-bridge knot.

19. Find vectors representing each of the rational tangles whose closure is one of the products of TN3 Resolvase.

20. Construct an example in which neither $A$ nor $B$ is rational and yet $N(A + B)$ is a two-bridge link with two components.

21. Prove that for any tangles $A$, $B$, and $C$, we have $N[A+(B+C)] = N[(A+B)+C]$.

22. Let $S$ and $T$ be tangles. Prove that we can obtain the knot or link $N(S + T)$ by gluing $S$ and $T$ together along the boundaries of their balls.

23. Suppose that $T$ is the trivial tangle $(0)$; prove that $N(S+T) = N(S)$. What happens if $T$ is the tangle $(0, 0)$?

24. Prove that the Whitehead link is topologically chiral as an unoriented link.

25. Suppose that in the statement of Theorem 7.1, the hypothesis that $N(S + R + R)$ is the figure eight knot were replaced by the hypothesis that $N(S + R + R)$ is the trefoil knot. Could we still prove that $R$ is rational, and that if $S$ is not rational then both $R + R$ and $T$ must be rational? Why or why not?

# References

Alexander, J. *The combinatorial theory of complexes*, Ann. Math. **31** (1930), 292–320.

Bates, A. D., A. Maxwell, *DNA Topology*, Oxford University Press, Oxford, England, 1993.

Boles, T. C., J. H. White, N.R. Cozzarelli, *Structure of plectonemically supercoiled DNA*, J. Mol. Bio. **213** (1990), 931–951.

Burde, G., H. Zieschang, *Knots*, W. De Gruyter, New York, 1985.

Buseck, P., S. J. Tsipursky, R. Hettich, *Fullerenes from the geological environment*, Science **257** (1992), 215–216.

Carina, R. F., C. Dietrich-Buchecker, J.-P. Sauvage, *Molecular composite knots*, J. Am. Chem. Soc. **118** (1996) 9110–9116.

Cerf, J. *Sur les Difféomorphismes de la Sphère de Dimension Trois* ($\Gamma_4 = 0$), Lecture Notes in Mathematics, Vol. 53, Springer-Verlag, Berlin, 1968.

Cerf, C. *Signed nullification number, remaining writhe, and chirality of alternating links*, J. Knot Theory Ramif. **6** (1997) 621–632.

Chambron, J.-C., J.-P. Sauvage, K. Mislow, *A chemically achiral molecule with no rigid achiral presentations*, J. Am. Chem. Soc. **114** (1997), 9558–9559.

Chen, C.-T, P. Gantzel, J. S. Siegel, K. K. Baldridge, R. B. English, D. M. Ho, *Synthesis and structure of the nanodimensional multicyclophane "Kuratowski Cyclophane," an achiral molecule with non-planar $K_{3,3}$ topology*, Angew. Chem. Int. Ed. Eng. **34**, (1995), 2657–2660.

Clayton, D. A., J. Vinograd, *Circular dimer and catenate form of mitochondrial DNA in human leukaemic leucocytes*, Nature **216** (1967), 652–657.

Conway, J. *An enumeration of knots and links and some of their related properties*, Computational Problems in Abstract Algebra, (Proc. Conf. Oxford 1967), Pergammon, Oxford, 1970, pp. 329–358.

Conway, J., C. McA Gordon, *Knots and links in spatial graphs*, J. Graph Theory **7** (1983), 445–453.

Crowell, R. H. *Genus of alternating link types*, Ann. Math. **69** (1959), 258–275.

Culler, M., C. McA. Gordon, J. Luecke, P. Shalen, *Dehn surgery on knots*, Ann. Math. **125** (1987), 237–300.

Dehn, M. *Die beiden Kleeblattschlingen*, Math. Ann. **75** (1914), 402–413.

Dietrich-Buchecker, C., J.-P. Sauvage, *A synthetic molecular trefoil knot*, Angew. Chem. Int. Ed. Engl. **28** (1989), 189–192.

Doll, H., J. Hoste, *A tabulation of oriented links*, Math. Compu. **57** (1991), 747–761.

Du, S., N. Seeman, *Synthesis of a DNA knot containing both positive and negative nodes*, J. Am. Chem. Soc. **114** (1992), 9652–9655.

233

234 References

Ernst, C., D. W. Sumners, *A calculus for rational tangles: applications to DNA recombination*, Math Proc. Camb. Phil. Soc. **108** (1990), 489–515.

Ernst, C., D. W. Sumners, *Solving tangle equations arising in a DNA recombination model*, Math Proc. Camb. Phil. Soc. **126** (1999), 23–36.

Flapan, E. *Rigid and non-rigid achirality*, Pacific J. Math. **129** (1987), 57–66.

Flapan, E. *Symmetries of Möbius ladders*, Math. Ann. **283** (1989), 271–283.

Flapan, E. *Rigidity of graph symmetries in the 3-sphere*, J. Knot Theory Rami. **4** (1995), 373–388.

Flapan, E., B. Forcum, *Intrinsic chirality of triple-layered naphthalenophane and related graphs*, J. Math. Chem. **24** (1998), 379–388.

Flapan, E., N. Seeman, *A topological rubber glove obtained from a synthetic single-stranded DNA molecule*, J. Chem. Soc. Chem. Commun. **22** (1995), 2249–2250.

Flapan, E., N. Weaver, *Intrinsic chirality of complete graphs*, Proc. Am. Math. Soc. **115** (1992), 233–236.

Flapan, E., N. Weaver, *Intrinsic chirality of 3-connected graphs*, J. Combin. Theory B **68** (1996), 223–232.

Freyd, P., D. Yetter, J. Hoste, W. Lickorish, K. Millett, A. Ocneanu, *A new polynomial invariant of knots and links*, Bull. Am. Math. Soc. **12** (1985) 239–246.

Frisch, H., E. Wasserman, *Chemical topology*, J. Am. Chem. Soc. **83** (1961), 3789–3795.

Graf, E., J.-M Lehn, *Synthesis and cryptate complexes of a spheroidal macrotricyclic ligand with octahedrotetrahedral coordination*, J. Am. Chem. Soc. **97** (1975), 5022–5024.

Hartley, R. *Knots with free period*, Can. J. Math. **33** (1981), 91–102.

Hoste, J., M. Thistlethwaite, *Knotscape 1.0*, http://www.math.utl.edu/morwen/knotscape.html (1998).

Hoste, J., M. Thistlethwaite, J. Weeks, *The first 1,701,936 knots*, Math. Intell. **20: 4** (1998), 33–48.

Hudson, B., J. Vinograd, *Catenated circular DNA molecules in HeLa cell mitochondria*, Nature **216** (1967), 647–652.

Jaco, W., P. Shalen, *Seifert fibred spaces in 3-manifolds*, Mem. Amer. Math. Soc. **220**, (1979).

Johannson, K. *Homotopy Equivalences of 3-Manifolds with Boundaries*, Lecture Notes in Mathematics, Vol. 761, Springer-Verlag, New York, Berlin, Heidelberg, 1979.

Jones, V. F. R. *A polynomial invariant for knots via Neumann algebras*, Bull. Am. Math. Soc. **89** (1985), 103–111.

Kanenobu, T. *Infinitely many knots with the same polynomial*, Proc. Am. Math. Soc. **97** (1986), 158–161.

Kauffman, L. *Formal Knot Theory*, Math Notes, Vol. 30, Princeton University Press, Princeton, NJ, 1983.

Kauffman, L. *State models and the Jones polynomial*, Topology **26** (1987), 395–407.

Kauffman, L. *Invariants of graphs in three-space*, Trans. Am. Math. Soc. **311** (1989), 697–710.

Kawauchi, A. *The invertibility problem on amphicheiral excellent knots*, Proc. Jpn. Acad. Series A **55** (1979), 399–402.

Kelvin, W. T. *Baltimore Lectures on Molecular Dynamics and the Wave Theory of Light*, C. J. Clay, London, 1904.

Kinoshita, S. *On elementary ideals of θ-curves in the 3-sphere*, Pacific J. Math. **42** (1972), 89–98.

Kohara, T., S. Suzuki, *Some remarks on knots and links in spatial graphs*, Knots 90, Walter de Gruyter, New York, 1992, pp. 435–445.

Krasnow, M. A., A. Stasiak, S. J. Spengler, F. Dean, T. Koller, N. R. Cozzarelli, *Determination of the absolute handedness of knots and catenanes of DNA*, Nature **304** (1983), 559–560.

Krätschmer, W., L. D. Lamb, K. Fostiripoulos, D. R. Huffman, *Solid $C_{60}$: a new form of carbon*, Nature **347** (1990), 354–358.

Kuratowski, K. *Sur le problème des courbes gauches en topologie*, Fund. Math. **15** (1930), 271–283.

Laird, T. *Development chemistry at its best*, Chem. Ind. **12** (June 14, 1989), 366–367.

Liang, C., K. Mislow, *Classification of topologically chiral molecules*, J. Math. Chem. **15** (1994a), 245–260.

Liang, C., K. Mislow, *Topological chirality of proteins*, J. Am. Chem. Soc. **116** (1994b), 3588–3592.

Liang, C., K. Mislow, *Topological features in proteins: knots and links*, J. Am. Chem. Soc. **117** (1995a), 4201–4213.

Liang, C., K. Mislow, *Topological chirality and achirality of links*, J. Math. Chem. **18** (1995b), 1–24.

Liang, C., K. Mislow, E. Flapan, *Amphicheiral links with odd crossing number*, J. Knot Theory Ramif. **7** (1998), 87–91.

Liu, L. L., R. E. Depew, J. C. Wang, *Knotted single-stranded DNA rings: a novel topological isomer of circular single-stranded DNA formed by treatment with Escherichia coli w protein*, J. Mol. Bio. **106** (1976), 439–452.

Liu, J., H. J. Dai, J. H. Hafner, D. T. Colbert, R. E. Smalley, *Fullerene 'crop circles,'* Nature **385** (1997), 780–781.

Livesay, G. R. *Involutions with two fixed points on the three-sphere*, Ann. Math. **78** (1963), 582–593.

Massey, W. S. *A basic course in algebraic topology*, Graduate Texts in Mathematics, Vol. 127, Springer-Verlag, New York, 1991.

Menasco, W. *Closed incompressible surfaces in alternating knot and link complements*, Topology **23** (1984), 37–44.

Mislow, K., R. Bolstad, *Molecular dissymmetry and optical inactivity*, J. Am. Chem. Soc. **77** (1955), 6712–6713.

Mitchell, D. K., J.-P. Sauvage, *A topologically chiral [2]-catenane*, Angew. Chem. Int. Ed. **27** (1988), 930–931.

Montesinos, J. *Variedades de Seifert que son recubridores cíclicos ramificados de dos hojes*, Bol. Soc. Mat. Mexicana **18** (1973), 1–32.

Morgan, J., H. Bass, eds., *The Smith Conjecture*, Academic Press, Orlando, FL, 1984.

Moser, L. *Elementary surgery along a torus knot*, Pacific J. Math. **38** (1971), 737–745.

Mostow, G. *Strong rigidity for locally symmetric spaces*, Annals of Mathematics Studies, Vol. 78, Princeton University Press, Princeton, NJ, 1973.

Motwani, R., A. Raghunathan, H. Saran, *Constructive results from graph minors: linkless embeddings*, 29th Annual Symposium on Foundations of Computer Science, IEEE, 1988, pp. 398–409.

Munkres, J. R. *Topology: a first course*, Prentice Hall, Englewood Cliffs, NJ, 1975.

Murasugi, K. *On periodic knots*, Comment. Math. Helv. **46** (1971), 162–174.

Murasugi, K. *Jones polynomials and classical conjectures in knot theory*, Topology **26** (1987a), 187–194.

Murasugi, K. *Jones polynomials and classical conjectures in knot theory II*, Math. Proc. Camb. Phil. Soc. **102** (1987b), 317–318.

236     *References*

Murasugi, K. *Knot theory and its applications*, Birkhäuser, Boston, 1993.
Nakazaki, M. *The synthesis and stereochemistry of organic molecules with high symmetry*, Topics in Stereochemistry, E. I. Eliel, S. H. Wilen, N. L. Allinger, eds., Vol. 15, Wiley, New York, 1984, pp. 199–251.
Nakazaki, M., K. Yamamoto, S. Tanaka, *Synthesis of [8][8] and [8][10]paracyclophanes*, Tetrahedron Lett. **5** (1971), 341–344.
Nierengarten, J.-F., C. Dietrich-Buchecker, J.-P. Sauvage, *Synthesis of a doubly interlocked [2]-catenane*, J. Am. Chem. Soc. **116** (1994), 375–376.
Ochiai, M., S. Yamada, *KnotTheorybyComputer, 3.60*, http://archives.math.utk.edu/II/software/mac/topology (1992).
Otsubo, T., S. Mizogami, Y. Sakata, S. Misumi, *Layered cyclophanes*, J. Chem. Soc. Chem. Commun. **13** (1971), 678–678.
Otsubo, T., F. Ogura, S. Misumi, *Triple-layered [2.2] naphthalenophane*, Tetrahedron Lett. **24** (1983), 4851–4854.
Paquette, L. A., M. Vazeux, *Threefold transannular epoxide cyclization synthesis of a heterocyclic $C_{17}$–hexaquinane*, Tetrahedron Lett. **22** (1981), 291–294.
Pasteur, L. *Recherche sur les relations qui peuvent exister entre la forme cristalline, la composition chimique et le sense de la polarisation rotatoir*, Annal. Chimie Physiques **24** (1848), 442–459.
Przytycki, J. H., P. Traczyk, *Invariants of links of Conway type*, Kobe J. Math. **4** (1987), 115–139.
Reidemeister, K. *Elementare begründung der knotentheorie*, Abh. Math. Sem. Hamburg **5** (1926), 24–32.
Robertson, N., P. Seymour, R. Thomas, *Sach's linkless embedding conjecture*, J. Combin. Theory B **64** (1995), 185–227.
Rolfsen, D. *Knots and Links*, Publish or Perish Press, Berkeley, 1976.
Rubinstein, J. H. *Heegaard splittings and a theorem of Livesay*, Proc. Am. Math. Soc. **60** (1976), 317–320.
Sachs, H. *On a spatial analogue of Kuratowski's theorem on planar graphs-an open problem*, Graph Theory, Lagow 1981, M. Borowiecki, J. W. Kennedy, M. M. Syslo, eds. Lecture Notes in Mathematics, Vol. 1018, Springer-Verlag, Berlin-Heidelberg, 1983, pp. 230–241.
Schubert, H. *Knoten mit zwei Brücken*, Math. Z. **76** (1956), 133–170.
Seeman, N. *Synthetic DNA topology, Molecular Catenanes, Rotaxanes and Knots*, J.-P Sauvage, C. Dietrich-Buchecker, eds. Wiley-VCH, Weinheim, Germany, 1999, pp. 323–356.
Simmons, H. E., III, J. E. Maggio, *Synthesis of the first topologically non-planar molecule*, Tetrahedron Lett. **22** (1981), 287–290.
Simon, J. *Topological chirality of certain molecules*, Topology **25** (1986), 229–235.
Simon, J. *A topological approach to the stereochemistry of nonrigid molecules*, Graph Theory and Topology in Chemistry, R. B. King, D .H. Rouvray, eds., Elsevier, Amsterdam, 1987, pp. 43–75.
Smith, P. A. *Transformations of finite period II*, Ann. Math. **40** (1939), 690–711.
Sritana-Anant, Y., T. J. Seiders, J. S. Siegel, *Design of novel aromatics using the Loschmidt Replacement on Graphs*, Carbon Rich Compounds I, A. de Meijere, ed., Topics in Current Chemistry, Vol. 196, Springer-Verlag, Berlin, 1998, pp. 1–43.
Suber, P. Knots on the web, http://www.earlham.edu/~peters/knotlink.htm.
Tait, P. G. *On knots I, II, III*, Scientific Papers Vol. I, Cambridge University Press, London, 1898, pp. 273–347.

Thall, E. *When drug molecules look in the mirror*, J. Chem. **73** (1996), 481–484.

Thistlethwaite, M. *A spanning tree expansion of the Jones polynomial*, Topology **26** (1987), 187–194.

Thistlethwaite, M. *Kauffman's polynomial and alternating links*, Topology **27** (1988), 311–318.

Thurston, W. *Three-dimensional manifolds, Kleinian groups and hyperbolic geometry*, Bull. Am. Math. Soc. **6** (1982), 357–381.

Van Gulick, N. *Theoretical aspects of the linked ring problem*, New J. Chem. **17** (1993), 645–653.

Walba, D. *Stereochemical topology*, Chemical Applications of Topology and Graph Theory, R. B. King, ed., Studies in Physical and Theoretical Chemistry, Vol. 28, Elsevier, Amsterdam, 1983, pp. 17–32.

Walba, D. *Topological stereochemistry*, Tetrahedron **41** (1985), 3161–3212.

Walba, D. *A topological hierarchy of molecular chirality and other tidbits in topological stereochemistry*, New Developments in Molecular Chirality, P. G. Mezey, ed., Kluwer, Dordrecht, 1991, pp. 119–129.

Walba, D., J. Armstrong, A. Perry, R. Richards, T. Homan, R. Haltiwanger, *The Thyme Polyethers "An approach to the synthesis of a molecular knotted ring,"* Tetrahedron **42** (1986), 1883–1894.

Walba, D., R. Richards, R. C. Haltiwanger, *Total synthesis of the first molecular Möbius strip*, J. Am. Chem. Soc. **104** (1982), 3219–3221.

Waldhausen, F. *On irreducible 3-manifolds which are sufficiently large*, Ann. of Math. **87** (1968), 56–88.

Wasserman, S. A., J. M. Dungan, N. R. Cozzarelli, *Discovery of a predicted DNA knot substantiates a model for site-specific recombination*, Science **229** (July 12, 1985), 171–174.

White, J. H. *Self-linking and the Gauss integral in higher dimensions*, Am. J. Math. **91** (1969), 693–728.

White, J. H. *Winding the double helix*, Calculating the Secrets of Life, E. Lander, M. Waterman, eds., National Academy Press, Washington, DC, 1995, pp. 153–178.

White, J. H., W. R. Bauer, *Calculation of the twist and the writhe for representative models of DNA*, J. Mol. Bio. **189** (1986), 329–341.

White, J. H., W. R. Bauer, *Applications of the twist difference to DNA structural analysis*, Proc. Nat. Acad. Sci. USA **85** (1988a), 772–776.

White, J. H., N. R. Cozzarelli, W. R. Bauer, *Helical repeat and linking number of surface wrapped DNA*, Science **241** (1988b), 323–327.

Yamada, S. *An invariant of spatial graphs*, J. Graph Theory **13** (1989), 537–551.

Yau, S.-T., W. H. Meeks III, *The equivariant loop theorem for three-dimensional manifolds and a review of the existence theorems for minimal surfaces*, The Smith Conjecture, J. Morgan, H. Bass, eds., Academic Press, Orlando, FL, 1984, pp. 153–163.

Yokota, Y. *Topological invariants of graphs in 3-space*, Topology **35** (1996), 77–87.

# Index